BOUND
TO THE
EARTH

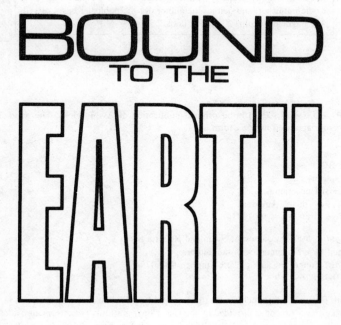

BOUND TO THE EARTH

JAMES A. SWAN, Ph.D. and ROBERTA SWAN

AVON BOOKS ◆ NEW YORK

BOUND TO THE EARTH is an original publication of Avon Books. This work has never before appeared in book form.

AVON BOOKS
A division of
The Hearst Corporation
1350 Avenue of the Americas
New York, New York 10019

Copyright © 1994 by James A. Swan and Roberta Swan
Published by arrangement with the authors
Library of Congress Catalog Card Number: 93-50764
ISBN: 0-380-76971-9

Library of Congress Cataloging in Publication Data:
Swan, James A.
    Bound to the earth / James A. Swan and Roberta Swan.
        p.   cm.
    Includes bibliographical references.
        1. Human ecology—Philosophy.   2. Philosophy of nature.
    3. Environmental degradation.   4. Conservation of natural resources.
    5. Renewable energy resources.   I. Swan, Roberta.   II. Title.
GF21.S93   1994   93-50764
363.7—dc20   CIP

First Avon Books Trade Printing: July 1994

AVON TRADEMARK REG. U.S. PAT. OFF. AND IN OTHER COUNTRIES, MARCA REGISTRADA, HECHO EN U.S.A.

Printed in the U.S.A.

OPM   10   9   8   7   6   5   4   3   2   1

TABLE OF CONTENTS

INTRODUCTION: LESSONS FROM THE PAST

Fifteen thousand years ago a naked man with a tattoo of a snake on his arm carries glowing embers carefully in the hollow of a stone. He senses the direction of the wind, chooses his spot, and starts a fire. As the fire sweeps across a meadow the man watches anxiously until the fire line reaches the trees at the ends of the meadow then sputters and dies. He smiles, happy that his judgment is good. The meadow is full of dead stubble, but the surrounding forest is moist and would take more than a small grass fire to make it burn.

Burning fields was probably the earliest form of land management. Humans used it to alter the natural succession of the plants for their own benefit. The people of fifteen thousand years ago knew a lot about ecology. They knew that the large grazing animals liked grass, but that the grasslands over time became forests. So, they systematically burned the grasslands to keep them open and have more abundant game to hunt.

In this early period of human intervention into nature little damage was done to the environment, but it was the start of our transformation of the earth. We were not doing anything much different than an alligator creating a place to wallow in the mud, or a beaver cutting down a tree, and, since our numbers were few, our impact was light. But, as our numbers increased, our transformation of the earth became more and more intrusive and destructive.

Ten thousand years ago we became farmers, and began to destroy or take over the habitat of other plants and animals who share the earth with us. We eventually gained so much control over nature's resources that we began to see ourselves as the masters of nature instead of part of nature. It is a delusion from which we are just emerging.

Although many of our scientific efforts are still geared to the mastering of nature, a new attitude is forming. This new attitude—actually a very old attitude, just lost for ten thousand years— says that we can be more successful by learning from and cooperating with the strategies that nature has already devised. We must, in effect, re-learn to copy nature's ways rather than try to circumvent them.

Nature has created a three-billion-year-old success story of life on earth. This success is reflected in the way that life is organized to make the best use of the resources available. Nature does not need to strip-mine to produce energy; nor are toxic wastes a by-product of its creation.

As we study nature's systems we are discovering more and more about how they function. We are learning that as the basic cells of life developed, they evolved a cooperative structure that initiated a chain of development leading ultimately to ourselves. And, we know that life spread around the globe to make the best use of each place it happens to be. And, finally, we know that plant life evolved to make the most efficient use of its surrounding resources.

This book does not attempt to give a definite description of how nature works, but rather explores how we can look to nature for ways to survive in a time of ecological crisis. We need to develop a way of thinking that is preventative and restorative. We can come to that way of thinking by looking at nature to reorient our thinking from a mechanical to an organic model. This is a book that suggests a path, not a detailed description of all you can discover. We have much to learn from nature.

In the first part of the book we look to the past and trace the way life is organized from its earliest forms, and we find that a basic theme for this organization is cooperation. We then follow life as it begins to multiply and cover the earth, and see how life organized itself into a larger pattern of cooperation, forming niches where each kind of life could make a living without constant competition. Finally, we step back and look at the larger themes of organization in nature, and find that pursuing the best use of the resources available is a lesson that nature teaches all around us.

We then look at the human past to see how human evolution evidences these same themes—and where the paths of humans and nature diverged.

In the second part of the book we take a practical look at how we can imitate the wisdom of nature in the present. We look at specific functions of nature, such as photosynthesis, to see how technology and human behavior can follow the patterns by which nature thrives. In this part of the book we will examine energy production, recycling and reuse of materials, manufacturing, efficiency, and transportation systems.

In the final part of the book we look to the future, at how the world might appear in the near future if we follow nature's examples in our work and personal lives. Our use of stories in this section follows the model of ancient cultures that traditionally used them to pass on information and to allow the listener to enter into another reality. The goal of the stories, just as in traditional tribal stories, is to jog your imagination, helping you see a new way to interact with the rest of the world.

We now use too many resources for too few people without considering the long-term results of our actions. We have fouled the air and the water; we have mined the land without restoring the lost nutrients under the guise of farming; we have created poisonous chemicals without any idea of how they will affect our children and great-grandchildren. And, we do it all in the name of "progress," thinking we were only working for the good of humanity.

Everyday the negative results of our actions are clearer and clearer. We must make changes. But we don't always know how. Polls show that the American public wants the environment protected. We are ready for change, if only someone will show us the way.

Perhaps the most hopeful aspect of the mess we are in is that we made it. It's not as though nature has a vendetta against human beings, or that there is some cosmic plan to rid the universe of our species. We created all the problems that we now have to solve.

This book is an attempt to give some insight into how that vast task might be accomplished.

As we explore some of nature's wisdom and try to reorient our thinking about our relationship with nature, we need to remember that the forest and the trees were there long before humans, long before any ancestor of our species was beginning to use rock scrapers and make changes in the environment. As we look to nature, we ask for the ability to find some of its wisdom and use it to create the next stage of human evolution.

The naked man with the tattoo of the snake on his arms walks back to his camp and is greeted by laughing children and barking dogs. He goes to the edge of the clearing and finds a basket with grain, collected by a woman of the village. He takes a handful of grain out of the basket and throws it into the air, offering thanks to the gods

that the field burning went well. Now, we, his
descendants fifteen thousand years later, offer
prayers for guidance as we attempt to understand the wisdom of nature and apply it to our
lives.

part I

ROOTS AND BRANCHES: UNDERLYING THEMES IN NATURE

The human mind and spirit are always striving to understand the wholeness of life. It is true that predation, violence, and death surround us, and that the law of "eat or be eaten" is also fundamental to nature. But these dramatic aspects of

nature are balanced by another side that mani-
fests in harmony and beauty. Now it is time to
look closely at this other side, focusing on ways
in nature which can suggest working models for
a transition to a sustainable society, a society that
not only provides for our needs now, but one
that will provide for the needs of our grandchildren and great-grandchildren.

1.

Nature at Work

F og pushes along the sides of Mt. Tamalpais near the Pacific Ocean in California. Tucked in a fold in the hillside below the mountain are the remnants of the giant redwood forest that once stretched from San Francisco Bay to the Oregon mountain ranges. The summer fog places a cooling blanket over the tops of the three-hundred-foot high redwood trees, and in the dry summer they absorb the fog and use it to keep alive and healthy until the winter rains come. This patch of redwood forest is part of an ancient climax community.

We like to take early morning walks in the ancient redwood forest, preserved in the Muir Woods National Monument, which is almost in our backyard. Muir Woods is visited by hundreds of thousands of tourists yearly, but if you go early enough, you can be the first humans of the day to visit the giant redwoods. In the moist, scented early morning air, deer are browsing on the young fern heads in the shade of the trees and on grasses growing in the occasional patch of sunlight. The birds proclaim their territory. An occasional squirrel runs chattering up the side of the tree while the fragrant, heady perfume of the woods fills your consciousness, making you wonder about your origins and ultimate place in the ecosystem.

The beauty of the woods inspires us, but perhaps

more inspiring is nature's amazing complexity in the
way it works. In the redwood groves we can find examples of how cooperation works in nature, of niches
that organize the economy of nature's systems, and
of nature's striving to make the best use of the resources available in the process of creating life's most
mature and complex development, the climax community, where an equilibrium of resource use allows
life to thrive without excess or waste. These are the
concepts we will be examining individually in this
chapter, but first let's see how they work together in
the redwood forest.

When you stand within the circle of Cathedral
Grove, with soft filtered light diffusing down through
the upper branches, it feels like you are in a medieval
cathedral. Your eyes are drawn up to the tops of the
trees; you wonder at their majesty as the treetops circle above and almost seem to join together. But there
is more than mystery and wonder here: the grove is
circular for a reason. Redwoods have a remarkably
shallow root system relative to their towering height.
Many mature redwood trees grow in circular groves,
roots intertwining to form a system of mutual support. Even those trees not in groves send their roots
far out to each side, hoping to link with others for
support. Without this cooperation between the redwood giants, high winds of winter storms could top-
ple them to the ground.

The redwood forest is like a giant apartment complex of many floors full of niches. Woodpeckers drill
holes in the bark halfway up the trunks. Crows, rav-
ens, and hawks roost in treetops. Jays nest on the
lower branches. Underfoot fungi live in the root system, their fruiting heads erupting in response to moisture. Tree squirrels live at the level of the
woodpeckers, scurrying to the floor for the seeds and

nuts that mice also relish. Ground squirrels and gray foxes burrow among the roots. All of these animals make a living in the woods, with minimal competition among them, as they work in their own niches.

A huge fallen tree currently blocks one part of the walkway through the woods. The rangers decided not to move it, but to let it return to the soil where it lay. None of the nutrients in the log will be wasted. Their release in decay will allow a luxuriant carpet of plants to spring up in the sunlight opening. Ferns and horsetails will thrive for a while in the open space, until a new redwood starts there; then, in a few years the ferns and horsetails will diminish in the shade of the new tree. Death and rebirth in a climax community works to keep the forest renewed, yet the same.

The redwood forest illustrates our themes of cooperation, creation of niches, and the formation of climax communities. It is the result of the harmony of nature's parts working together, supporting each other to reach their ultimate expression of beauty and efficiency.

As we leave the redwood forest, our journey through nature begins as we go back in time and look closely at the inner working of the cell, where the theme of cooperation reveals itself as a prime organizing force in nature.

COOPERATION: THE CORE OF ALL LIVING ORGANISMS

Cooperation is not a concept we usually apply to nature. We are more apt to think of predation and competition. But cooperation is the balancing act that allows predation and competition to exist.

As we examine the inner workings of the early

cells—the only form of life that existed for two billion years—we find this early life survived through a basic cooperative structure. The original bacterialike cells cooperated by trading, or exchanging, selected parts of DNA with each other. Trading DNA is like having access to instant genetic engineering. Modern scientists were quite surprised when they found out that DNA swapping, which they have worked so hard to perfect, has been freely used by bacteria for billions of years. DNA exchange allowed bacteria to respond quickly to environmental changes. Frequent volcanic eruptions, meteors bombarding the earth, a blazing sun four times as hot as it is today were all conditions that created a fluctuation in climate and demanded swift adaptation of the single-celled organisms.

After the first billion years some single cells moved to the next stage of evolution characterized by "symbiosis," which happens when two separate organisms join together for mutual benefit to meet a new goal, or to cope with an adverse situation in a new and creative way. Sometimes organisms join together permanently as we'll see; others remain separate, but entirely dependent, on each other. Several simple bacterialike cells, for example, permanently joined together to form a new kind of cell called the nucleated cell.

Just as its name implies, the nucleated cell has a nucleus, a place deep inside where the DNA has extra protection in the form of an additional membrane. This cell, unlike the bacteria, doesn't share its DNA. But what this cell does do is take in boarders, who at some point became permanent residents. One boarder specializes in breathing oxygen, while another provides locomotion, and finally—in plant, not animal cells—there is a boarder that uses sunlight to create energy. Cellular symbiotic cooperation joins the strengths of different organisms together for mutual benefit, a pattern of cooperation. Symbiosis is repeated throughout nature, inspiring us to attempt to design more cooperative human systems.

For example, algae and fungi join together to produce an entirely new life form, a lichen. Although lichen can be found in almost everyone's backyard, they also grow in harsh environments like the Arctic tundra. The algae within the fungi's crusty, gray-green cover is able to capture enough sunlight to provide food by photosynthesis; it then feeds the fungi. The fungi in return give structural protection to the algae, and anchor it to the stone. But their joining is so complete that they are no longer separable. This is a permanent symbiosis, as in the new cell.

Symbiosis also exists between organisms that maintain separate structures and lives. In the ocean "cleaner" fish actually swim into the mouths of sharks and barracuda to eat the parasites that accumulate there in the teeth and gums. In the African plains, birds called ox pickers land on the backs of water buffalo and eat off the parasites. The animal is healthier and the bird has its dinner. And back in the ocean, some crabs form symbiotic relationships with sea anemones. The anemones permanently attach themselves to the crab's claws and sting would-be attackers. Thus the crab is protected and the anemones get a wider feeding range than if they were stuck to a rock.

As we look around in nature we find even more examples of symbiosis. Cows cannot digest their food without the bacteria in their stomach. And flowering plants could not exist without the symbiotic help of bees and other insects. All these variations of symbiosis derive from the initial symbiotic pattern developed in the nucleated cell. This pattern of cooperation and symbiosis ultimately led to the development of multicelled organisms like trees and people where billions of cells cooperate.

Once cells cooperate and join together to form larger organisms, these larger organisms need to survive in relationship to one another. One of the ways they do this is by evolving, or finding, a niche, the next stop on our journey through nature's past.

NICHES IN NATURE: EVERYTHING HAS ITS PLACE

When we look quickly at natural systems the order is not readily apparent. Birds scatter in seemingly random patterns among the trees; animals run in all directions; weeds appear without invitation or reason.

But the birds, the animals, and the plants all occupy their own special places in nature. We call these places "niches." The word niche was taken by ecologists from the vocabulary of architects where it means a small place in a church where a statue would sit: a special place for the images of gods and saints.

When cells joined together and formed more complex life forms, they began to develop endless new strategies for making a living on different parts of the earth. Each new organism had to find its niche. Moving from the oceans, where life evolved in the shallow protected waters, to the land about four hundred million years ago, microscopic plants began to colonize the land. Blue-green algae fertilized the first landscape, then mosses and lichens covered the land surface.

Life exploded into a wondrous variety of new forms when plants and animals moved onto the land. Mosses and lichens, followed by the club mosses and the seed ferns, were just the beginning in the long succession of plants that has led to our current profusion of plant and animal species. New plant forms helped improve the soil and provided habitats to be used by the insects and worms, the first animal life to inhabit the earth.

Much literature of the early twentieth century describing the evolution of life emphasizes the idea that nature in her evolution was ruthlessly combative: as followers of early evolutionary biologists might say, it was "red in tooth and claw." But in the later part of the twentieth century it has become also clear that the same principles of cooperation that guided the early evolution of the cell continued to operate as a balance to competition when nature became more complex. Rather than competing, different forms of plant and animal life look for places where their own special characteristics are best used. These are the ecological niches.

Plants look for sunlight, since they all photosynthesize—make their own food from sunlight and water—but, in order to use all the different places on earth where sunlight is available, different strategies to access that sunlight have evolved. Imagine that plant life had only evolved in outright competition. All the plants would try to become taller and wider, continually occupying more land, and reaching above the other plants in direct competition for the sunlight. But that is not what has happened. Instead, plants range in size from the microscopic algae to the giant sequoia, living underneath, on top of, and beside each other, each seeking a unique way to make the best use of nature's resources.

Plants have also adapted to take advantage of specific local climates. In the Arctic, in the desert, in the humid tropics, plants adapt themselves to local conditions and find places to grow and strategies to propagate themselves. Succulents like the cactus in the desert have evolved skins with few pores, which help the plant to hold water for the long dry times between rains. Cactus seeds will only germinate when they sense moisture from a drought-breaking rain. In the temperate forests of North America the jack pine has cones that will not open until it has been through a forest fire and it is only through forest fires that the underbrush is sufficiently cleared out to allow space and sunlight for the new pine tree to sprout.

Niches define how an organism lives. A fern would dry up in the desert heat, but is excellent at converting low-level sunlight into sugar where there is adequate moisture. The cactus would not grow in the dim cool light in the understory of the redwood forest, but thrives in the sizzling desert heat that would kill the fern.

Animals occupy niches as well. From the insects on the forest floor to the largest predator, animals specialize in what they eat and where they eat it. Grazing animals eat different plants, so that, for example, in and around the Serengeti plain in Africa, as the vast herds move across the plain, the wildebeests and the gazelles eat the grasses and the giraffes eat the tops of the acacia tree. Predators—lions, jackals, and hyenas—follow and eat the weak and young of the grazing animals, serving their purpose by keeping the numbers of herd animals in check. Finally, our relatives the chimpanzees, who eat both meat and plants, live on the edges where they have access to both the forest and the plains eating insects, small animals, fruits, and nuts.

Even populations of similar birds that seem to intermingle in the trees will stay at different levels to reduce competition with each other. Flycatcher birds live in the upper branches of a tree where they catch insects in the air, while robins live in the lower branches and feed on the forest floor. When an animal leaves its own niche and attempts to intrude on another, conflict can erupt, but, even then, most conflict is solved without bloodshed by mutual threats of aggression, leading to the weaker species backing away. You could easily observe this behavior around suburban garbage cans when a raccoon and possum, both garbage can scavengers, confront each other. After mutual threats of aggression the possum will slip away, leaving the spoils of the evening to the raccoon.

When we know where to look, the niches of plants and animals are easy to find, with plants and animals adapted to use the resources of their niche to the fullest advantage. But there is a scale of relationship beyond niches, where groups of plants and animals cooperate in creating nature's most efficient use of energy, in the climax community.

NATURE'S SEARCH FOR PERFECTION: THE DEVELOPMENT OF THE CLIMAX COMMUNITY

When the plants and animals in an area have reached the point of using only as much energy as they create, they have become a climax community. This marvelous state of balance and equilibrium can be seen in mature ecosystems around the globe. A forest, a prairie, or a tundra can all be climax communities. In the Arctic plains where it is too cold and windy for trees to grow, a community of mosses and lichen reach a balance where life and death are equal, and yearly growth is the same as yearly decay. On the prairie, with variable rainfall patterns and grazing herd animals, a rich community of grasses renews itself every year, taking the nutrients it needs from the soil and putting back an equal amount. And in forests with adequate rainfall and rich soil the same type of ultimate balance in nutrient cycling is achieved.

A climax community in nature can be a model for humans in their search for sustainable development. The climax community is the end of a long process in nature called "succession," where during the process plants produce excess growth and nourishment. Succession begins with "pioneer" plants, continues through a variable number of middle stages of plant life, and ends in a self-sustaining, ongoing climax community.

To better understand how a climax community is generated let's take a detailed look at how the climax forest renews itself when it is damaged.

The forest is damaged as a blast of cold wind rushes down from the Arctic and turns a winter rainstorm in the Midwest into a blanket of ice. Burdened by the weight of the ice, branches of the weakened trees in the mature oak-hickory forest begin snapping off and crashing to the forest floor.

Then final destruction comes in spring when a towering black thundercloud looms up, promising rain, but instead shoots lightning bolts into the trees. The giant trees ignite in flames and a few hours later, only a few scattered blackened trunks and piles of ashes remain.

Now, the forest will seek to recreate itself in the pattern of succession. The first splotches of green will be grasses, quickly taking hold and thriving in the nutrient-rich ashes from the fire. The grasses will send down roots, holding the soil in place, preventing erosion. As blades of grass die and are replaced by new ones, the organic content of the soil will increase and so will its water-holding capacity.

Seeds of annual flowers, blown in by the wind or carried by birds and animals, find these conditions suitable to their growth and add roots to help hold moisture. Soon the green carpet of grass will be decorated with colorful flowers. Quail and meadowlarks nest in the grasses and field mice tunnels are everywhere. The birds keep the insect population in balance, and the mice eat excess seeds and help to fertilize the forest. Shrubs like hawthorne, dogwood, and viburnum find their way into this ecosystem and begin to flourish. Deer eat the shrubs and grasses.

Passing squirrels bury acorns, hickory nuts, and walnuts. Soon young trees begin to raise their heads above the shrubs in their quest for sunlight. The first trees are sun-loving, short-lived aspens and cottonwoods. Next come the more slowly growing oaks, hickories, and pines.

Without sunlight, many of the original pioneer plants die off, eliminating the habitat for some animals. Short-legged field mice, best suited for tunneling through grass, disappear to be replaced by long-legged wood and white-footed mice who can run rapidly and climb trees. Squirrels and raccoons, who nest in trees, replace groundhogs who never leave the ground.

Thirty years after the original fire, an oak-hickory climax forest occupies the site. Its re-creation into a climax community has been accomplished by the succession of plants and animals that conditioned the soil by adding nutrients. Finally, a state was reached where the soil was fertile enough to support the oak and hickory trees and a balance was achieved. The oak and hickory trees take what they need from the soil and return an equal amount of nutrients.

All parts of the process of succession leading to a climax community are important to nature and relevant to human communities. Human communities, like plant communities, build on earlier efforts as they search for ways to make the best use of the resources available. The climax community is a goal, once achieved by our ancestors, that we now seek to re-create in a sustainable society.

North America was once heavily covered by climax forests such as the oak-hickory forest. Today we only have small parts of that original forest scattered across the land—the rest have been sacrificed to lawns, parking lots, and farms. Where we do find the remains of the ancient, climax forests, we are drawn to their fragrant, shaded interiors, with little undergrowth where you can walk easily under the huge trees.

Cooperation, niches, and climax communities are the themes of nature we have considered in this chapter. Now, we will look at how these themes apply to the early development of human communities. Humans, who started out as unconscious participants in life's ongoing evolution, now need to consciously embrace the life-giving themes of nature and model their own attempts at creation on her designs. But before we look at how we might model our current endeavors on nature's designs, let's look at how evolution once helped humans find a perfect fit with nature and create the original human climax community.

2.
Tugging on
Nature's Web

The wild cards in natural systems are human beings. Our mutable species has been able to leave the boundaries that restrain the rest of nature. We have more choice than any other species. But originally our choice was tempered by our reverence for nature, and we generally acted within a spiritual covenant based on copying the ways of nature. When we began to attempt to alter nature's patterns with the development of agriculture ten thousand years ago, we still maintained a deep respect for the ways of natural systems. That respect gradually eroded as we gained more control of nature.

Our divergence from nature accelerated in the last two centuries when, propelled by science and technology, Western civilization lurched off in new directions, yielding tremendous developments. Now, we no longer mesh with nature's patterns, nor give it our reverent respect, and we are destroying the world around us. Modern humans are searching for new models to restore nature kinship before it is too late.

For ten thousand years we have been tugging on nature's web trying to shape it to fit our desires. And now, the web is beginning to fray and come apart in places. Can we find a place in nature where we fit with the other forms of life on earth rather than destroying them? We certainly can't become cave people again. We must find a new synthesis.

We have seen how nature cooperates, allowing life to flourish by creating niches, and making the best use of the resources available in climax communities. These abilities *are* also deeply imbedded in humans, as we will see in this chapter.

HUMANS AND COOPERATION

Humans, the same as all other life forms, are composed of cells. We have nucleated cells that are cells within cells, life within life. The symbiosis and cooperation that are primary underpinnings of the ecosystems of nature are part of every cell in our body.

But cooperation goes beyond our cellular structure; it has been necessary in our social organization because of our relatively weak bodies. Early humans were fangless, clawless creatures, neither herd animals nor solitary predators, who took care of their young in small kinship groups.

We base our picture of early human life on a few clues skillfully assembled. The pelvic structure from skeletons three million years ago shows that humans had already attained an upright posture. Standing upright demanded a stronger pelvis, which had a smaller opening for the birth canal. The structural change in the human pelvis helped create the long childhood of the human species. And this prolonged childhood, as much as the need for mutual defense, mandated the need for social cooperation.

For human infants to keep their relatively large brains, they must be born early enough in their development to fit through the birth canal while their heads are still small. As a result, human infants emerge as tiny vulnerable beings, not even strong enough to cling to their mothers as their chimpanzee cousins do.

For a minimum of five years human children, the most helpless in nature, must be watched and cared for before they are capable of developing any skills they can contribute back to the group that has provided the necessary food, clothing, and shelter.

Our limited knowledge of the prehistory of humans shows repeated behaviors that were cooperative. The hunting of large ice-age mammals had to be a cooperative effort—it takes more than one scrawny human to bring down a six-thousand-pound woolly mammoth! And, of course, once the animal was dead it had to be skinned, cut up, and preserved, or eaten before the meat began to spoil. This is an arduous process similar to what Eskimos do today when they kill a whale and the whole village turns out for the butchering. The building of large enclosures, such as the prehistoric one discovered at the fifteen-thousand-year-old site in Mezhirich in the Ukraine and made out of the bones of giant mammoths, could only have been done by an organized and cooperative group. And, finally, the wonderful cave paintings at Lasceaux in France required a group effort to bring in the materials and create the light necessary for the artist to work.

These ancestors were hunter/gatherers who lived by hunting wild animals and gathering roots, fruit, and nuts. The small hunter/gatherer groups that remain such as the Kung bushmen, African pygmy, and Australian aborigines still live in clan structures where the welfare of the whole group is considered in making any decision. Their tribal decisions and rituals are based not only on their cooperation with each other, but on what they see as cooperation with nature.

Historically tribal groups cooperate with the rhythms and cycles of nature in order to survive. Nature is their supermarket and its hours are unpredictable. Their sensitivity to the natural world is so great that they can smell a change in the weather, a hint of moisture in the air telling that rain is coming, or a trace of sand or dust telling a sandstorm is coming, and a displaced stone speaks of game.

Hunter/gatherers learned the plants, the animals, the cycles of weather, and the soil conditions in their territory through a tradition that stretches back three million years. They knew that cooperation with each other and with nature was their best hope for an abundant life.

HUMAN COOPERATION IN THE LAST TEN THOUSAND YEARS

As human communities have grown larger, cooperation in groups has become more fragmented. Today we live in one place, work in another, send our children to school in another. We may create support networks, but these networks are not the people we would necessarily be with if a hurricane, earthquake, flood, or urban violence struck where we live or work.

Mobility in America has fragmented patterns of social cooperation. We might fly across the country for holidays with family members that we love, but who are not part of our everyday reality. We move in and out of neighborhoods an average of every five years. And, we change careers on an average of three times in a lifetime. It's a world of change that leaves us with a craving for old-fashioned communities, but little direction on how to get there. Communication technologies link us with others, creating a sort of electronic community, but the physical intimacy of community is a less common event in modern life.

Although cooperation is not embedded in our lives as it was when we were hunter/gatherers, the voice of cooperation in our genetic structure is waiting under the surface, ready to come out in times of crisis or natural disaster. When floods devastated the Midwest in 1993, people came from great distances to fill sandbags and try to stop the rising waters. In the San Francisco earthquake of 1989 people who didn't know each other or the possible victims in the buildings risked their lives trying to find survivors in the rubble. The same thing happened in the Mexico City earthquake of 1991.

The desire is there, now we need modern cultural structures to encourage more cooperative behavior to be expressed and supported. The desperation of this need is underscored by the increasing violence across America where teenagers without the limitations of traditional community support and restraints fill our emergency rooms with gunshot victims nightly.

Not only do we need to develop cooperative structures between humans, we are also learning that to survive as a species we must also yield and cooperate with nature. For example, this kind of cooperation might happen in a more organic approach to farming where we reject pesticides and artificial fertilizers and concentrate our research on copying nature to create healthy plants.

One of the most significant ways that humans co-operate is in creating ways to make a living; in other words, to create their niche.

THE HUMAN NICHE: WHERE DO WE FIT IN?

Niches exist as a means of allowing all parts of nature to thrive without spending all of their energy on competition. Humans originally made their living as hunters/gatherers, that was the human niche.

The niche early man occupied was that of the hunter/gatherer at the top of the food chain. (Because of our diet and hunting habits our closest niche relative is the bear. We both are omnivorous hunter/gatherers, able to eat a wide variety of foods and requiring a large territory in which to survive on the land.)

We were able to occupy this niche because of our cooperative social structure. In hunter/gatherer societies all members of the group take care of each other. The men hunt and the women forage for fruits and other plant foods. Most of the hunting and gathering is done in groups. In some tribes, the hunter who has made the kill is the last to receive his portion, since it is the duty of the hunter to provide for the others. And if there is any food at all it is shared. We defended ourselves by banding together, using stones and sticks for defense and for hunting.

We foraged as we moved through the countryside, gathering roots, nuts, and fruit. Insects were probably a favorite food as they still are with many tribal groups, and we rounded off our diet with fish and the flesh of small-to-medium-sized animals.

As our level of social cooperation and our skills at making tools increased, we learned to hunt bigger animals, to store food, to make clothes, to build shelters, and to communicate more and more successfully with each other. But we were still hunter/gatherers.

The food that hunter/gatherers found differed according to where they lived. Inuit in the Arctic hunted caribou, whales, seals, and caught fish. Since fruits and vegetables are in limited supply in the Arctic, some may have been dried for winter. But to obtain most of the needed vitamins and minerals the Inuit ate all parts of the animals they hunted: organs, blood, and bone marrow.

In the temperate zones of the earth, hunter/gatherers would stalk the herd animals, gather the wild fruits and grains, and dig roots when necessary. With each season there were different food opportunities available such as the annual run of the salmon and the return of ducks and geese. The roots were always there, but they preferred meat when they could get it.

And in the tropics, the rain forest tribes still in Africa and South America have the most fresh fruit and root plants available. They eat small game when they can catch it, but also spend their time hunting beehives for honey and picking apart termite mounds for tasty grubs.

At the top of the food chain, with a social organization allowing us to lend strength to each other and a large brain allowing us to make and use weapons, we evolved methods to hunt anything that the largest predator could. During our tenure in our hunter/gatherer niche, where we thrived for 99 percent of human history, we spread around the globe, inhabiting every continent and most islands. Then, ten thousand years ago we began to leave the hunter/gatherer niche in the most profound shift in lifestyle in human history.

BEYOND THE HUNTER/GATHERER NICHE

We left the hunter/gatherer niche when we became serious in the development of agriculture. In that new pattern, farms transformed the landscape. Where we had shared the earth with all other creatures, we now began to create places where they were excluded. But farming before the industrial revolution, no matter how widespread, did not seriously encroach on the niches of other creatures.

Without machinery our transformation of the landscape was slow, allowing our fellow creatures to adapt. The most harm done to nature without machinery was when we tried to farm the marginal lands like hillsides and caused erosion of the soil, or when our flocks of sheep or goats ate the grasses, shrubs, and trees down to the bare roots, again allowing soil erosion to occur. Even these destructive patterns evolved over time and, for example, the denuded hillsides of the Mediterranean became the sparse but sustaining olive and lemon culture of travel brochures. Our real impact on the environment began when we created machines to do our work for us.

With machines we could use the products of the earth faster. Mining, machine-based agriculture and the growth of cities began to displace more and more of our fellow creatures. The ratio of humans to their surrounding environment became like that of insects. Beehives and termite mounds are the only structures in nature that have the density of our large cities.

In the past fifty years our impact on the earth has dramatically escalated. The efficiency of our machines has allowed us to use resources at a dizzying rate.

Our farms cover thousands of acres, forcing out the wildlife that lived in the tree and brush cover between farms. In our cities and suburban areas we pave over the ground, eliminating the forests and killing the bacteria in the soil. Our manufacturing processes poison the ground water and leave piles of toxic waste. Species are being eliminated at a pace equal only to the great extinctions of the geological past, such as the disappearance of the dinosaurs.

Because of the scale of everything we do, we are eliminating diversity in nature. But nature loves diversity. Monocultures are far more easily devastated than diverse ecosystems. We do not know what the consequences will be if we remove too much of that diversity, but the laws of ecology suggest that the fewer species there are in an ecosystem, the more susceptible it is to damage and destruction.

We now face a question of overwhelming importance: How do we as humans define our new niche? We are a species whose occupancy of the earth is increasing in geometric proportions, without any idea of how this will affect our future or the planet. Technology has given us the hubris that we can always find a new device to fix any need or problem that arises. Recognition of the web of life, that all things are ultimately interconnected, has waned.

We are creating a fool's paradise, seemingly solid, but poised on the brink of massive failure when we finally eradicate one too many of our fellow creatures. We once knew how to live in harmony on the earth; we were part of a natural climax community. Perhaps if we look at how we once functioned as a climax community and try to understand what caused its demise, we can get some ideas on how to start to create a new sustainable niche for the human species.

THE HUMAN SEARCH FOR THE BEST USE OF RESOURCES: WHEN HUMANS EXISTED IN A CLIMAX COMMUNITY

Native Americans would say that humans should walk in balance on the earth, take what we need, and give back an equivalent amount. Their description of balance is what happens in a climax community. The hunter/gatherer community that we were part of was a climax community and had been one for several million years.

What exactly do we mean when we say hunter/gatherers operated as a climax community? Basic to our definition of a climax community is that a balance of energy and material use has occurred. For human hunter/gatherers creating this balance required not only a knowledge of the plants and animals that they lived with, but it also required a philosophical viewpoint, a firm belief in the necessity of their conserving actions.

The hunter/gatherers knew every tree and rock and bush in their territory. They knew that they could only take part of a plant if they wanted to have more of that plant next year. They knew where the water sources were, and were not only conserving in their use of them but regarded most water sources as being inhabited by spiritual forces.

Plants could not be gathered, nor could animals be hunted, without the proper prayers and without being in the correct frame of mind. When respect was shown, the people believed animals would sacrifice some of their species in exchange for long-term vows of reverent stewardship. Their conserving behavior was based on an attitude of reverence toward the environment in which they lived, and an understanding that to have renewable resources available every year they must deliberately limit their use of resources.

If we examine hunter/gatherers as a climax community in order to get ideas of how we can create a sustainable society for ourselves, it would be valuable to see what this kind of community looked like. How did they decide the proper amount of resources to take from their habitat, and how did they express their reverence toward nature?

These tribes, whom we assume had much in common with tribal cultures surviving into the twentieth century, had a lifestyle many would envy. They seldom worked more than three or four hours a day; leisure time was spent in storytelling, games, and performing religious rituals. Stress levels were very low. They were secure that nature would provide, as long as they respected and followed its laws, which they taught in the songs, dance, art, and stories of their tribe.

Australian aborigines, American Indians, and numerous African tribes survived as hunter/gatherers into the twentieth century. Their ways of hunting and finding food, their population density, and their attitudes toward the environment are all in keeping with the basic principles of a climax community—that the resources used must be in constant balance with the resources being created.

The characteristics of these peoples which define them as climax communities are their steady state populations, their use of resources where they never take more of any plant or animal than can replenish itself, and their attitudes of reverence toward nature.

Australian aborigines, for example, tell in their myths of the time in the past when humans and the rest of nature were one, and even the rocks were alive to them. They call it the "dream time." And now, when they hunt or gather plants, they believe that the plants and animals, in essence, their ancestors, have willingly come to offer themselves to be eaten. Therefore, they cannot waste anything that is offered so generously. Their relationship to everything in nature is based on this belief in being part of the interrelatedness of all living things.

This belief in the unity of all life is typical of hunter/gatherer traditions extant around the world. It is a belief system in harmony with the conservation of resources in a climax community.

But most hunter/gatherers gave up their way of life. They learned how to herd the animals as well as hunt them, and they learned how to grow food as well as gather it. Agriculture and animal husbandry slowly led them out of a climax community that had endured for centuries. If we examine what we know about how hunter/gatherers lost their place in nature, we find they destabilized their climax community and moved into what in ecology is referred to as a disclimax community: a community in which the climax condition has been disturbed, like a forest after a large fire.

But why did we develop agriculture? It may have been for the ease and reliability of having plants in our own backyards rather than searching the countryside for them, just as we began to keep and breed animals so that they were readily available rather than hunting them. Since the rise of agriculture seems to coincide with the end of the last ice age, it seems possible that a change in plant and animal species available, due to climate change, may have caused an increase in planting that had been going on sporadically for some time. But agriculture itself was not the reason that we were no longer a climax community.

How did our ancestors move away from the climax community they had as hunter/gatherers? The creation of agriculture and the development of industrial technology may seem like the obvious answer, but the real issue is the size of population relative to available resources.

As hunter/gatherers we learned to control our population relative to our resources. We knew if we overstepped the boundaries of population growth eventually starvation would result. Agriculture and the creation of surplus food allowed us to step over those boundaries.

Agriculture enabled us to allow our population to rise right to the level of our ability to feed everyone in good times. When droughts and other bad times came there wasn't enough food, and masses starved or suffered.

We had established a new pattern for determining how many people were in the world. We would have as many children as the best of times would feed, instead of keeping populations down to what we could accommodate in the worst of times as we had as hunter/gatherers. Populations rose to the level of subsistence living where large groups of people were never quite satisfied but survived and reproduced.

Population increased slowly until the Industrial Revolution when it began to explode. The Industrial Revolution created machines, transportation systems, and chemicals that could expand the food supply. But this did not lead us to wealth and abundance for everyone. Instead, the population stayed at subsistence level for most of the world. More food just seemed to create more babies.

Although at this time pockets of abundance exist, how long they can continue to exist seems to be a question we must now face more realistically.

We are no longer a climax community. We take in more resources than we put back, and we need the resources of more and more other species just to exist. We have become like nothing else in nature, a creature whose cleverness allowed us to multiply out of balance, appropriating for ourselves the land, water, and air that had been occupied by other species, and living on borrowed time.

Perhaps one of our greatest losses from leaving the hunter/gatherer climax community is that loss of reverence for nature which allows us to understand and live by nature's laws. We are no longer sure of what is right and wrong in our actions concerning nature. We turned our backs on the hunter/gatherer way of life, and left paradise behind. As a dis-climax society we may never find our way back to paradise. Can we at least find a way to walk in balance and stop our destruction of the earth?

Should we allow our population to increase until we have taken over all the places of the other large animals? Should we deliberately limit the human population so that other species may co-exist? Would we be saving ourselves by allowing for the diversity nature needs to survive? Or are such man-made controls on man also interfering with nature? More than ever before, answering these ethical questions will affect the fate of the entire species.

We Were Meant to Live in Balance

Now we face the greatest challenge our mammalian cleverness has ever encountered: Can we bring ourselves into some kind of balance and harmony with the natural resources available to us? There is little doubt that we have the intelligence to figure out the situation. And there is little doubt that we have the ability to rapidly deploy new social systems and new modes of technology.

The biggest barrier we face to creating a massive change is the lack of awareness that we need to make a change. Not that scientists haven't tried to warn us. Thomas Malthus, an English scientist contemporary with Charles Darwin, warned us of the possibility of exponential population growth. Conservationists like George Perkins Marsh, Theodore Roosevelt, Gifford Pinchot, and John Muir began to sound warnings of the destruction of the wilderness a century ago. In 1947, author and environmentalist Aldo Leopold warned us of the destruction of whole ecosystems and called for a new ethical belief system to help us recover our natural niche. And for the past thirty years ecologists have warned us of the destruction that has been occurring because of our increasing population and use of resources. And certainly we have become more aware of environmental damage and are taking steps to stop it. But simply putting your old newspapers out by the curb is not enough to reverse the damage we have caused. We must look at our activities as whole systems, and find all the places where we are responsible for environmental denegration. Then we must change those activities.

But in spite of our wonderful large brains we still tend to react to threats only if they are perceived to be immediate. Perhaps that is understandable. For the first three million years of humanoid existence our technology and our numbers were not sufficient to make an irreversible change in the pattern of nature. In evolutionary terms this power is new. Now, following the example of the most ancient bacterialike cells we must adapt quickly or possibly become extinct.

As we strive for a new climax community we need to embrace the underlying principles of nature that made the hunter/gatherers successful for so long. We need to make the most efficient use of resources available and leave enough surplus for bad times. And we need to cultivate a new attitude of reverence for the earth.

Twentieth century civilization has produced mountains of garbage, like "Mt. Trashmore" near Detroit, which is used by skiers in the winter, or whole new islands created off the coast of Japan. Common sense tells us that we cannot keep taking raw resources and turning them into garbage forever. The raw resources will run out and we will be left with mountains of trash instead of trees and grass and animals and water. Could the mountains of trash be a reflection of our current values and mental attitudes?

A human climax community is one in which the output of goods and services is equal to the energy returned to the system by the recycling and reusing of those goods and services; it is a system that sustains itself. Any object created, a car, a house, a pair of roller skates, takes raw materials and energy. In a climax community none of the raw material is ever thrown away. It is returned to the system to be used again, and the energy is created by renewable sources.

Nature has evolved systems over billions of years that work in harmony with each other, that build from bare, rocky, thin soil to lush, green forests. Without human intervention the processes of nature have evolved self-regulating forces of beauty, grace, and efficiency. Our challenge is to learn how to honor them and be inspired by their truth to create new cultural values and systems.

MOVING FORWARD, LOOKING BACKWARD

Humans exist at the pleasure of natural processes. It is the power of the sun that creates warmth, and the work of bacteria and worms that makes the soil fertile. And we are creatures of the earth just as all the other animals are: dependent on sun, soil, and water for our very life. And we are now creating our new niche, for ourselves, our children, and many generations in the future. Can we now take nature's lessons and apply them to create a climax community?

After looking at the themes of cooperation, niches, and climax communities in nature and in human societies we have a sense of how deeply imbedded these concepts are in us. As we now turn to look at practical applications from specific examples of nature's processes, we hope to devise ways of making a living on the earth more in tune with nature's rhythms. It is time to think about repairing the frayed edges of nature's web.

part II

HOW TO THINK LIKE A TREE

As we try to repair the broken links on nature's web, we need models for new ways to create energy, food, and transportation.

To help us find these models we will travel inside a tree, crawl on the forest floor, ride on the wings of flying seeds, and explore the web of life as seen in whole ecosystems, so that we can isolate helpful examples of how nature creates sustainable life. The tree, the forest floor, and the seeds will become our teachers as we plan how to redesign our world.

We will see that building our homes, businesses, and industries in ways that create a healthier planet will help us create happier, healthier, more fulfilling lives that are also economically sound.

In this part of the book we will concentrate on how specific problems in our world can be fixed. And, not forgetting our philosophical reasons for making these changes, our awareness will never be far from our overarching theme that nature is the inspiration and model to help us create a sustainable society.

We will show how the themes of cooperation, niches, and climax communities apply to these practical concerns of constructing our environments. Cooperation with the process of nature is a theme in chapters on energy and transportation. We see the advantages of niches when we discuss industrial ecology. And the lessons of a climax community clearly apply when we discuss recycling, manufacturing, and efficiency.

3.

Emerald Alchemy
from the Sun

The creation of energy demands a large proportion
of the resources we wrest from the earth. This is a process
where cooperation with the forces of nature is essential.
Plant life in its evolution developed in cooperation with
the sunlight that falls on the earth daily. As we look at
how plants already work with the sun and how we might
imitate their efforts, think of how this organic joining of
human technology to nature's way can help us in our
path to a sustainable community.

High on a mountain in the southeastern desert of California an ancient tree thrusts its bare, gnarled branches into the air. The tree, a bristle cone pine, estimated to be four thousand years old, is related to the earliest cone-bearing trees on earth from three hundred million years ago. It exists only in these mountains where harsh, cold winds blow so hard that trunks of successful trees must grow parallel to the ground.

This tough and tenacious plant survives by growing only very slowly, and storing energy from the sun inside its green needles, thus continuing the work that all plants do everyday—performing photosynthesis.

Photosynthesis, as you probably remember from high school, is the process through which sunlight strikes the cell, excites the molecules, and creates energy in the form of sugar, which nourishes the plant. The familiar mysteries of photosynthesis are well worth reviewing since they contain seeds of possibility we have yet to apply to our technology.

All food on earth begins with photosynthesis, making humans and animals completely dependent on plants for creating food. Only plants can do it, and if they did not photosynthesize the whole food chain would collapse and we would starve. Photosynthesis not only occurs on the land, it also occurs in the water where a mass of food is produced by the photoplankton and blue-green algae that form the base of the food chain in a water environment. We seem to have forgotten this basic lesson when we go around the world destroying the green cover of the earth and poisoning the life in the sea and other waterways.

We learn such basic concepts as creating food through sunlight in childhood, but seem to forget their true meaning as adults. Food doesn't grow in supermarkets, and energy is transformed—but not created—by the local utility company. Food grows as the sun's incredible life-giving energy, the source of all energy on earth, pours down on the planet.

As we look to nature for guidelines and inspiration, the use of energy from the sun is a good place to start. Presently we depend heavily on fossil fuels for energy, using up nonrenewable resources and creating monumental air and water pollution. We could have a much cleaner planet if we took a lesson from the plants about how to most effectively create energy. There are two aspects of the sun's energy we want to look at: energy for photosynthesis and direct heat that is stored in rocks, water, and biomass.

Of the many possible ways to collect energy from the sun, photovoltaics, a modern process similar to photosynthesis, is one that could solve many of our current energy problems. Photovoltaics, usually referred to as PV, is a process that creates electricity by capturing the electrons from sunlight to produce electrical energy.

The process of photovoltaics was discovered in 1840 by Alexandre-Edmond Becquerel, a French scientist. He found that electrical currents could be produced by some chemical reactions that were stimulated by light. Although this information was noted by the scientists of his time, it was not until well after the turn of the century that serious research was conducted on this phenomena.

In 1953 a team of scientists working at Bell Laboratories created a cell that was able to take sunlight in and generate electricity. Solar cells capture the energy in sunlight using the photovoltaic process and converting it to electricity which is then either used directly, or more commonly, stored in a battery to be used as needed. Once we perfect ways of directly using, or collecting and storing, this energy the supply is renewable as long as our sun lives, since only 1 percent of the sunlight that falls on the earth daily, if captured as solar energy, could produce far more electricity than is generated by all other presently known sources combined.

It's easy to see the similarity between solar powered electricity and photosynthesis. Just as the sugar produced by photosynthesis is the lifeblood of the tree, electricity has become the lifeblood of our industrial society. Both seek to harvest the energy that pours directly onto the earth everyday from the sun.

If every roof on every structure were covered with photovoltaic material and if photovoltaic power were combined with the lower energy needs of increasingly energy-efficient appliances, lighting, and machinery, photovoltaics could come close to meeting the energy needs of much of the world.

Until the mid 1970s solar cells were expensive and painstakingly handcrafted. But the energy crisis and the higher prices of Middle Eastern oil in the 1970s changed that. By the late 1970s research and development funds were being allocated by the government and private industry into the further development of the solar cell. Development brought the price down, but solar-cell-generated electricity still costs five to ten times as much as utility-produced electricity. However, the oil-based energy crisis helped turn around much of the resistance to solar energy, thanks to an important factor: reliability of supply. Solar cells, although still expensive, began to be seen in a new light.

Solar cells were already being used in remote areas where it was prohibitively expensive to run utility lines for electricity. But some pioneering personalities in the 1970s began to have visions of houses powered by solar cells that made them independent of the primary power grid. To such people the cost of the system was less important than freedom from what they considered an unstable electrical source. Being "off the grid" was the dream of many people who were part of the back-to-the-land movement of the 1960s and the 1970s. It was a small group that had walked away from the dreams of mainstream America. They fine-tuned their storage batteries and charge controllers, and generally disappeared from America's mainstream consciousness for twenty years.

But many people, such as John Schaeffer, founder of Real Goods, a company that sells alternative energy products, continued research and sales. Real Goods was founded in 1978 with a vision that never faltered. People would need the products and services of non-fossil-fueled renewable energy sources, and Real Goods would be there when people were ready.

Located in Willits, California, Real Goods has now started the Institute for Independent Living, where people who are ready to "go off the grid" can come to weekend or week-long classes to learn the fundamentals of what they need, how to assemble it, and where to buy it. At the end of the class they go shopping for their solar panels, inverters, and batteries at the Real Goods showroom. The Institute for Independent Living is a blend of the 1990s: idealism and practicality coming together. It is both a great marketing tool for Real Goods and a real service to the people who want to develop a new way of supplying themselves with electricity.

A counterpart to Real Goods is the Sun Electric Company founded by Daniel Brandborg in 1984 in Montana. Since then the company has installed over one thousand of the estimated hundred thousand homes in this country operating off the grid. Crafts people, retirees, and celebrities roam the hills of Montana living in various degrees of comfort in solar homes designed by Sun Electric. With six photovoltaic panels and propane cooking you can run an energy-efficient household or put on forty panels and run a regular high-energy consumption suburban house. Sun Electric has done them all.

Cheaper land prices and the desire for privacy drive some people to live far away from utility lines. Gas- and diesel-driven generators have provided electricity for many people in these locations for decades, but the noise of a generator, the odors, and the fumes are unwanted intrusions in a pristine natural setting. RV owners who frequently use generators are

also turning to small portable solar arrays, panels that
they can put on top of their RVs, for trips away from
the campgrounds where they usually get their electricity.

One of the most unique "off the grid" homes is
actually a boat. George McNeir III has designed a solar electric canal boat. This is a luxury craft with all
the comforts tucked into a thirty-by-eleven-foot space.
Solar cells on the boat's roof feed the storage batteries
that power household electricity and the electric motors that run the boat. Designed for cruising the
twenty-five thousand miles of inland waterways of
America, Mr. McNeir sees his boat as an alternative
to the energy-intensive RVs so beloved by retirees.
Totally silent in its operation and energy independent,
this beautiful craft fits the vacation fantasies of more
than just retirees.

In order to integrate a new technology into our way
of life we need information on how it works and examples of where it has been used successfully. In 1989
the town of Willits, California started a Solar Energy
Expo and Rally as a way to showcase solar energy.
The Expo has demonstrations on solar hot-water heating where you can take a sun shower, on how photovoltaic panels work where you can figure out the
number of batteries needed to run your solar array,
and on related wind and hydro technologies that
round out your solar energy package. You can see all
the equipment that you need for solar living and find
out how much human muscle it takes to store energy
for watching your VCR in the evening with a bicycle-
driven generator. But the big draw is the race by solar
and electric vehicles: the electrathon.

The electrathon is a race of homemade vehicles, like
a soap-box derby, except these vehicles are all "elec-
trically assisted." These spunky little vehicles range

from slightly modified bicycle frames to sleek aero-dynamically groomed fiberglass, cigar-shaped tubes. Their shapes range from the practical to the exotic, suggesting that getting around town in the solar future could be a lot of fun.

Solar and electric car races provide a showcase for new techniques and experimental models. The longest and most challenging solar car race is held in Australia, where cars from around the world race over the flat, sunbaked outback for eighteen hundred miles. This race is taken as a serious challenge by all those who participate. For example, the 1992 University of Michigan winning entrant, the Sunrunner, was worked on by 120 students during more than a year of fifteen-hour days. To do this project they raised one million dollars from more than twenty corporations. The thirty-five competitors in the 1992 race were from nine countries and included all the major automobile makers and solar cell manufacturers.

In Switzerland one thousand electric cars with batteries powered by solar systems are now used for transportation. The official government goal is to have two hundred thousand solar electric cars on the road by 2010.

These elegant solar race cars won't be on your city streets, but by the turn of the century you will find an increasing number of electric vehicles. California has passed pollution reduction laws which will guarantee solar and electric cars on the road by the year 2000. In Sacramento the Sacramento Municipal Utility District has built the first solar charging station consisting of eight large solar panels for their demonstration fleet of electric vehicles and is building another station at the Sacramento airport.

Widespread use of solar power depends on cheap solar cells and increasing battery efficiency. Current limitations on electric vehicles include the size of batteries, their relatively short life span, and the long recharging time. Major American and Japanese auto makers are racing to develop a more efficient battery.

In attempting to bring the price of solar-generated electricity down to competitive levels Texas Instruments and Southern California Edison have undertaken a joint venture to develop and manufacture solar cells. This is the largest effort to date in the United States, and they hope to cut the cost of the basic production of solar cells by 90 percent.

Texas Instruments has been manufacturing semiconductors, the material for computer chips and solar cells, since 1959. Now they have a new process, discovered in the early 1980s, which promises to bring the price of solar-created electricity down to what is currently paid for conventional generation.

Utilities have not always been enthusiastic about photovoltaic power. At the same time that serious development of photovoltaic energy use was beginning, all the attention, money, and glory went to the development of nuclear power as the energy source of the future. Even satellites in space, which now rely solidly on PV cells, were scheduled in the original scenarios to be run by nuclear power.

The initial response of utilities to solar energy was to test the idea of "solar farms" where huge arrays of PV cells and mirrors were set up in the desert to create new power stations. Some solar energy will probably be used this way, but uses more economical and in keeping with creating new models that mimic nature's sustainability would be decentralized and individual. Utilities are finally starting to pick up on these. A utility, after all, provides electrical power, and is not tied to any special way of doing this.

Two utilities are starting to put PV modules on people's houses. The Idaho Power Company, instead of running expensive power lines, will buy a PV system for houses located in remote settings, then charge the homeowner rent for the system. And the Sacramento Municipal Utility District is putting PV units on the roofs of houses already served by power lines, and feeding the energy back into the main power grid. They see that in time, collectively, these small units could produce significant amounts of electricity, lessening the demands on large power plants.

Smaller uses of PV power, but important ones in making solar better known to the public, are creeping into daily use. Lights on pathways can be powered by sunlight during the day and reflect it back at night. The call boxes along the freeways easily store the small amount of electricity needed to make calls. Hydraulic pumps on farms and ranches run easily with small solar and battery systems and calculators run by solar cells are in many households already.

The energy from the sun can be captured in ways other than mock photosynthesis. Many household and manufacturing processes can be done just by collecting the heat of the sun. In nature plants and animals take advantage of sun-heated rocks, sand, and dirt. A heat-seeking plant grows along a rock face. Lizards, snakes, cats, and humans lie on sun-baked rocks, soaking up the energy the rock has gathered from the sun.

People have put this knowledge to work for thousands of years. There is evidence that caves with south-facing openings, naturally warmed by the sun, were preferred by Paleolithic hunters forty thousand years ago. More recently ancient pueblo construction in the American Southwest made use of a similar principle.

The pueblos built of adobe, a brick made of the local mud or clay, absorb heat during the daytime and release it at night. This works to both heat and cool the interior. In the winter, the sun strikes the walls at a low angle, allowing them to absorb the most heat energy available. By evening the heat has penetrated through the walls and is radiated to the interior. In the summer the angle of the sun causes the sun's rays to glance off the sides of the buildings and the pattern of use is reversed, with people sleeping on the roofs to get the cool evening air as the day's heat was seeping into the room, and working in the relative daytime coolness inside their adobe structure.

To show how universal the concept of stored and released heat is, we can look at ruins of ancient human villages and compare them with buildings still being constructed. Early in the 1960s James Mellert, an archeologist, found some of the oldest "urban" structures, occupied from 6500 BC to 5500 BC, in a place called Catal Huyuk in Turkey. The construction of these buildings was almost identical to the pueblos in New Mexico—halfway around the world and seven thousand years later. This kind of construction is still used in Turkey near Mellert's find and around the world in India, China, and the higher altitudes of South America.

Two thousand years ago at the height of the Roman Empire there were laws that protected people's rights to have access to the sun. In the troop garrisons and new Roman towns it was illegal for buildings to block each other's sunlight. In a small Roman city you might see houses oriented so that they received a maximum amount of the winter's sun. Even the trees were placed so as not to shade the buildings.

It is such a simple concept that it is a wonder that over time people seemed to forget it completely. The south sides of buildings stopped opening up to the sun, roof overhangs obscured its rays, and trees were allowed to block the winter sun's rays that would have been entering south-facing windows. As we replaced cooperation with nature and ignored natural forces, we created artificial air-conditioning in hermetically sealed buildings. Aside from using more energy resources than necessary, this also created the "sick building syndrome," where the inner atmosphere of a sealed building becomes toxic to its occupants from the out-gassing of fumes by all the synthetic materials used in its construction.

Methods of working with nature were all revived in the 1970s when America began to realize how fragile and dangerous their dependence on oil was, and began to consider what alternative sources of energy might be available. Using these ideas, once dismissed as idiosyncratic and not practical for large-scale use, developer Mike Corbett created a large-scale project, building a two-hundred-unit subdivision called Village Homes in Davis, California in the 1970s.

Mike Corbett is a developer with a vision: to create sustainable housing by cooperating with nature as much as possible. The winding streets and footpaths in Village Homes, lined with edible fruit trees and other foliage which also help control the climate, take advantage of solar energy before it reaches the houses. The streets are narrow and the footpaths are made of stone and have grass and moss growing in the spaces left between the rock, creating more greenery to help repel the outdoor heat in the wilting Davis summers.

The houses in Village Homes take advantage of solar heating and cooling principles by having maximum southern exposure, and tile or stone floors or walls to absorb the heat. They also have carefully designed roof lines so that the summer sun is reflected away from the structure. Plantings, frequently edible, are placed to avoid shading the southern exposures in the winter and to provide as much shade as possible in the summer. The styles vary but the beauty and serenity of a Mediterranean villa pervades the development twenty years later.

The use of solar energy is not just being picked up in trendy California. Soldiers Grove, Wisconsin (pop. 693), was founded in 1864 on a mud plain of the Kickapoo River and flooded six times in sixty years. Finally, it was relocated to higher ground in the late 1970s, and at that point the city passed an ordinance that all commercial establishments must derive at least 50 percent of their heat from the sun. Turk's IGA hasn't had to turn on its backup gas heater since it was built in 1980. Similar stories could be told about the Country Garden Restaurant on Sunbeam Road or the Schoville Office Building on Sun Hill Road. After twenty years people in Soldiers Grove have found solar energy not only saves money, but like Village Homes, they receive visitors from around the world who are looking for examples of sustainable development.

Heating our dwellings and places of work by the sun may be the biggest potential application of the direct use of solar heat in the US, but in developing countries around the world food preparation may be the most urgent use of direct applied solar heat. Many third-world countries are in tropical areas and use what little vegetation is available for cooking fires. This is leading to deforestation, which has serious consequences for the planet as a whole, as well as local ecosystems. A very simple solar device, a solar cooker, can help solve this problem.

Solar cookers are basically small boxes with glass tops that get hot enough inside to cook food. The temperature inside a solar cooker can reach two hundred degrees within a half hour and up to 350 degrees during midday sun conditions. A box might be made of a plywood outer shell with fiberglass insulation, a metal plate on the bottom painted black, aluminum foil sides, and a top made of glass with a reflecting panel.

The sun's rays hit the glass top and are trapped as heat inside. The effect is the same as being in a closed car on a hot summer day, only intensified.

The governments of China and India have begun small-scale distribution of solar cookers. But the real work is being done by an international network of small relief organizations like Solar Box International of Sacramento, California. They go into villages and try to adapt the solar box principle to local materials, family size, and cooking techniques. They say that requests for help and information from third-world countries have increased exponentially during the past two to three years due to a steady decrease of local cooking fuels.

The simple principle of trapped heat from the sun can cook food, preserve food, warm buildings, and save millions of dollars and millions of gallons of oil in the process by something as fundamental as putting large windows on the south side of a house and having some way to store part of the trapped heat inside. Once we start to work seriously with this principle, increased uses of the direct heat from the sun will be found in manufacturing and business as well as in the domestic sphere.

The compass plant, a small sunflower of the Great Plains which has leaves that are oriented to the four directions of the compass to maximize exposure to the sun, is a good teacher about solar energy. It moves continuously during the day, orienting itself so that it can receive maximum sunlight. Nature thrives using the warmth and energy of the sun, and even in the harshest conditions like that of the bristle cone pine, the life-generating process of photosynthesis occurs. The sun pours more energy on the earth every day than we have even imagined ways to use, and almost all of this energy is deflected back into space without our trying to capture it.

If we are to become a sustainable life form on the planet we quite obviously need to learn to capture the sun's energy by designing according to natural forces. Throwing away this gift every day is proof that our species has yet to understand how to live in harmony with nature. The power of the future, the sun, is waiting for us to discover as many imaginative ways as possible to capture it, and use it in sustainable ways for the development of all humankind.

4.

Leaves that Fall
Only to Rise Again

Total recycling of all materials used by human societies is an important step in leading us to a new climax community. We need to be inspired by the thought that we can understand how nature works when it recycles, and use our hard-gained scientific knowledge to create new ways to copy nature's methods more precisely.

In the fall leaves turn yellow or red and provide a spectacular blaze of glory before they fall to the ground, the colors showing the beauty of their understructure after the life-producing chlorophyll has gone. They are dead. But in nature the death of one element only contributes to the life of another. If this were not so living things would quickly suffocate in their own waste.

Humans have totally isolated themselves from nature's cycle; we *are* beginning to suffocate in our own wastes. We preserve the debris of modern life, our dead things, as garbage in landfills, just as we try to preserve our dead loved ones in lead-lined caskets which deny and only slow the inevitability of decay.

Nature constantly reuses her resources, and humans need to follow her example in everything they create. We need to think like trees. For nature, death releases nutrients and energy that will become new life since nature designed all life, even human bodies, to be recycled.

Green leaves on a tree are kept alive by a constant infusion of water, rising from the roots and up the trunk of the tree, until the fall when the tree shuts down its water supply to the leaves so that it can survive the winter without freezing. The leaf stem then loses its hold on the tree, and the wind shakes it to the ground where the dry, brown cellulose skeleton of the leaf is left to crunch under our feet. As the fall rains come, the miracle of death to new life begins.

Earthworms begin to feed on the leaves, and the parts of the leaf that the earthworm doesn't eat are attacked by bacteria and fungi, reducing the leaf to smaller and smaller particles. After being digested by worms and microbes the minerals released from the dead leaves seep into the soil. Encountering water, the nutrients dissolve and, in the spring, are drawn up into the trees, where they are available for the process of new growth. The undigested part of the leaf stays with the top layer of soil, increasing its depth and ability to hold water.

A similar process happens with any dead organic matter from all plants and animals on the earth. When an animal dies in the wild, large scavengers such as hyenas or vultures eat the meat they can wrest from the bone. What remains of the flesh, skin, and fur is attacked by smaller decomposers—insects, worms, and bacteria. Within a few months all that is left are the bones, which still have important nutrients, bleaching in the changing seasons. Minerals such as calcium phosphate are dissolved from the bones by rain, then broken down further by bacteria. This process releases calcium and phosphate back into the soil where they are used by the plants, which are then eaten by the animals and produce new bones.

In nature death provides the raw ingredients for each new cycle of life. All parts of dead plants and animals find their way back to life after the decay process has reduced them to usable nutrients absorbed by the soil in nature's elegant system of recycling.

Recycling has become a part of our daily vocabulary and simple recycling procedures are becoming commonplace. We don't always do it, but we know we should. More and more people put their papers, cans, and bottles out by the curbside or cart them away themselves. Curbside recycling is available in four thousand communities nationwide, and forty-one states have legislation on the books to move toward total recycling, many because their landfills are overflowing.

One reason that we need legislation is that in an amazingly short period of time, since the end of World War II, we expanded the old dump to a landfill, and packed the landfills to the brim. Large cities around the country cry that they have run out of landfill space, and many have. That doesn't mean that there isn't land left, but rather that no one wants a landfill in his backyard. Landfills, even with concrete liners, leak toxic chemicals into the groundwater, provide a breeding habitat for rats and bacterial species we would rather not encourage, and, of course, they stink.

Some people think burning garbage to get rid of it is the answer. That not only produces toxic ash, but it is not in keeping with the regenerative processes in nature, where what we create must be seen as a resource to be used and reused.

As we try to learn to produce less waste, we are

developing new techniques to handle what we have already created. In addition to recycling, we are beginning to handle large quantities of waste by composting and bioremediation, processes where specially introduced, and, sometimes artificially created, bacteria gobble up the contents of a landfill and render it harmless.

Composting is a way of speeding up nature's processes of decomposition. Alternating layers of soil and organic wastes are laid down in a bed called a compost pile where wastes are literally cooked, enabling the bacteria to do their normal work at a higher speed. As waste material is added and the process of decomposition begins, heat is generated by the gases that are naturally given off by decomposition, which are trapped in the pile and cause the temperature to rise.

Compost has to be tended by being turned, and by adding new material to keep the temperature right. Eventually the pile is "done." It is then free of all garbage odor; in fact it should smell like freshly dug earth, and is ready to be applied to the farm, garden, or yard.

Until recently composting was done by all farmers. But farmers began to use manufactured fertilizers, and abandoned composting, except for the few practicing "organic" farming who did not believe in using chemical fertilizers. Now organically produced foods sell for premium prices as customers begin to value the possible health advantages of food produced without pesticides or chemical fertilizers. But composting is no longer just for farmers; it is starting to help solve problems at our overflowing landfills.

Compost is now being made in industrial settings and municipal dumps around the country. Will Brinton and Clark Gregory, research biologists specializing in compost at Woods End in Maine, consult with clients from all over the world. They have made separate compost piles out of two hundred tons of crab shells and guts, old telephone books, and TNT-laced soil from a manufacturing plant.

The process of industrial and municipal dump composting is far more complex than what the farmer does. The right mixture of ingredients must be used. For example, to compost the TNT, Will Brinton used apple pomace, chicken manure, potato wastes, straw, and sawdust. With the right mixture of wastes all the items in the compost heap break down into their elemental compounds. Even the ink in newsprint gets 98 percent digested. Dynamite is digested and no longer explosive, and uranium's form is changed so that it precipitates out as water solution and can be collected separately.

Even heavy metals and toxic wastes may soon be transformed by composting. In the process called "bioremediation," toxic disposal specialists are experimenting with using bacteria to digest the offending substances and turn them back into something useful. The Homestead Mine in South Dakota is cleaning up one hundred years' worth of cyanide poisoning with bacteria found to eat cyanide. Groundwater Technology, Inc. in Massachusetts uses a bioremediation process in which natural bacteria consume benzene commonly found in abandoned gas stations and Superfund toxic sites, such as the Inger Oil Refinery in Mississippi. Oil spills, such as the massive one which dumped eleven million barrels of oil into Prince William Sound in Alaska, can also be partially cleaned up by bioremediation. Alpha Environmental of New Jersey has a patent on a microbe that eats oil, and is using it on oil spills around the world.

The successful use of composting and bioremediation on complex industrial wastes is leading some

people to think about what might be done in municipal dumps, and some areas of New York State are
beginning to experiment with composting material
that would ordinarily go to a landfill.

If you want to compost the leaves on your lawn and
your own household wastes the fastest way is using
worms—the same earthworms are already hard at
work in your lawn or garden. To learn how, get a
copy of *Worms Eat My Garbage* by Mary Appelhof.

It is also possible to safely compost human wastes.
Although most of us are happy flushing the toilet and
not thinking about where the wastes go, human
wastes can be composted and reused. Most human
wastes go into our sewage systems along with less
polluted water from showers, dishwashing, and
sometimes rain water. The relatively small volume of
human waste contaminates and vastly increases the
difficulty of treating the total volume of waste water.
This is a problem for several reasons: first, fresh water
is an increasingly scarce commodity due to population growth and an increase in the total amount of
water that households, industry, and agriculture use.
Second, polluted water must be treated and cleaned,
an expensive and energy-consuming process, before
it can be released into a natural setting. A composting
toilet deals with the waste product at its source, the
way nature does.

The composting toilet was invented and first used
in Scandinavia where cottages on sandy beaches
made outhouses or septic fields impractical. Heat, oxygen, and organic material transform human wastes
into usable compost. When we visited Norway in the
late 1960s we went to stay at a beach cottage. The
bathroom contained a large toilet unit that you had
to climb a stair to get to. The family we were visiting
told lots of jokes about their funny-looking toilet, but

there were no complaints. It was odor-free and did its job efficiently.

A composting toilet has a toilet bowl attached to a large rotating drum. The bottom of the bowl opens and drops the waste into the drum. Periodically you drop in kitchen wastes, moss, or other organic material to the drum, and turn the drum to mix the contents. After several months about half the contents of the drum are dropped into a drawer which is under the drum where the composting process is completed. After several months the waste has turned into compost.

A more sophisticated version of the composting toilet features a low flow toilet, using a pint of water per flush, with the composting drum outside and below the bathroom. This toilet uses electricity to provide heat to aid the composting process and evaporate any excess liquid.

Both systems depend on a venting system with negative-pressure air flow to take and keep odors away from the user area. Sometimes a small fan is used to insure that no backdraft occurs. But users of these systems are universally happy with their performance. Many people are amazed that the system can work without odor.

And like any successful form of composting, the final product is odor free and ready to apply to the garden. Since most composting toilets now are in rural areas, the compost can go directly in the ground. Maybe if this were allowed in the city a new version of the ancient oriental tradition of collecting "night soil" would develop, pre-processed and sanitized for our modern world. The compost could be collected and taken to nearby farms or parks.

Because they both save water and recycle valuable nutrients people have wanted to use composting toilets in places other than remote settings. However, most cities ban their use. City officials either don't believe that the composting process works or they don't trust that people could carry out the simple acts of adding kitchen wastes and turning the drum of the composting toilet. Critics of this ban have compared most cities' sewage process to taking clean drinking water, adding human wastes, using an expensive cleaning process to get rid of most of the human wastes, then sending the water back to people as drinking water. No wonder so many people are buying bottled drinking water!

People in America are now familiar with the recycling of paper, glass, metals, and plastic. The economics and technology for mass recycling of these products is just passing out of an awkward stage. In the early 1990s there were often surpluses of paper and glass as enthusiastic people tried to recycle. The markets were still immature. But in 1992 the Buy Recycled Business Alliance, a group of businesses dedicated to buying recycled products, was formed with twenty-five members. A year later the alliance had six hundred members and had spent $11 billion on recycled products. The organization publishes an official guide to recycled products which started with five hundred entries and now lists six thousand.

Even with the exploding growth in markets for recycled products there still may be collected recyclable material that is hard to sell, but it is usually in places far from manufacturing centers where transportation makes the price too high for the collectors to make a profit. And as new government programs go into effect, money should be available to start businesses in those areas to use the surplus material.

One-third of all the paper produced in the world today is used by Americans. It has been such a common part of our lives that we seldom treat it like the miraculous substance that it is. Paper consists of the long fibers that make up plants chopped into smaller parts and pressed together to make a flat surface. When you are reading your morning paper you are holding a piece of a Douglas fir in your hands. The shoe box you brought home last week is really a piece of a Southern pine forest. Very expensive papers are made from linen or cotton fibers and approach or exceed the durability of fabrics. We are surrounded by natural fibers which can be transformed into a substance to give us information, protect our food or clothing, help us transport products, or help us clean up spilled orange juice.

Plastics are finally being recycled after a long time of simply being ground up and reused. Plastics are sometimes difficult to recycle because of their mixed chemical formulas, but new manufacturing standards and new labeling on bottles has made sorting easier and recycling more possible. Coke and Pepsi are both using at least 25 percent recycled material in their PET plastic bottles. Other plastics are bought by handlers and reclaimers who prepare material for recycling including vinyl, polypropylene, and HDPE. They then sell the appropriate type back to manufacturers. When plastic is hard to separate into a specific formula, the mixed plastic is ground up into pellets, heated, and made into new forms like toy cars and plastic lumber.

We don't use nearly as much glass, domestically, as we used to, but when clean, sorted quantities are available for recycling, it is simply melted and used again. It is the simplest of all recycling processes, but is the one lagging most behind at the moment. Glass is only acceptable if it is color sorted and without any contaminating elements, such as metal or ceramics.

Much of the glass collected from home use is not in that category, which means that at the moment some of the glass people have carefully put out at the curbside is not always getting reused. It might end up in the landfill instead.

This is a time of transition where problems will occur as we move from looking at the world as a set of disposable materials strictly for human use to the world as an intricate web of interacting organisms of which we are one element. Part of the solution is to continue recycling, buy recycled products, and reuse items whenever we can.

We need to remember that as recently as fifty years ago trash was routinely sorted through, the usable items sold. Food waste, whenever possible, was composted or fed to animals. Our local waste collection agency is run by Joe Garbarino, who is the third generation to run the family business. Joe remembers clearly in his childhood riding on the trucks and sorting out eyeglasses, clothing, and anything else saleable that was being thrown away.

In fact, reuse, which we will define as keeping material goods that we have manufactured in circulation as long as possible, needs to be a major part of our strategy for creating a sustainable future. Where recycling can take as much energy to recreate a product as it did to make it in the first place, reuse takes little energy, just a lot of imagination and skill.

An item in this country greatly in need of reuse is tires. It is estimated that three billion tires exist in dumps around the country. One reason we have so many old tires is that we do not encourage the use of retreads on passenger automobiles. When the treads wear down, the tires are thrown out instead of being given new treads and used again. Retreads fell out of favor when new tires began to get sixty thousand to eighty thousand miles of use per tire instead of the twenty thousand when tires were being retreaded. To be strong enough to be used as retreads tire cores must be made stronger than they are now and that makes the original purchase more expensive. Trucks and heavy equipment still use retreads where the cores are made to stand five hundred thousand miles of use. This is an issue that definitely needs further attention since 90 percent of used tires are now being looked at just as something else to be incinerated.

A few creative people are finding uses for some of the overwhelming number of discarded tires. Actor Dennis Weaver has had a house constructed in New Mexico using tires as the structural element and covering them with adobe. Now at least another fifty houses in the nearby Rocky Mountain foothills are made of tires, and a whole company, Earthships, has been created to build "earthship houses," the name the designer has given these self-contained, low-energy-use houses. In certain areas, shredded tires are mixed with asphalt for road resurfacing; this new material reduces highway noise and has greater ice resistance than conventional asphalt paving.

And, although not likely to make a big dent in the 240 million tires produced per year for the nation's landfills, Used Rubber USA has creatively found a way to make purses, pouches, knapsacks, and sandals out of old rubber tires.

They get their materials from dumpsters and tire dealers. Then they take them home and use a lot of imagination and design training to cut, stitch, and fasten the rubber pieces together with rivets and washers. The finished product is waterproof, stainproof, and comes with a lifetime guarantee. Their shop in San Francisco is small but growing.

Other creative people around the country are starting businesses by reusing things we no longer want, like Heidi Gerquest of Maine who takes used materials of all kinds and makes them into furniture, or Richard Spenser of California who takes apart old machinery and reassembles it for new purposes, or Cornelia Kietzman of Vermont who takes used adult-sized clothing and makes clothes for children, or Jim Broadstreet, an architect, of Missouri who builds whole houses out of reused material. These people, the pioneers of a whole new movement, have taken items out of the waste stream and given them new life.

The term "waste stream," which refers to everything we throw away, needs to become obsolete if we are to become a sustainable society. All materials that we create or manufacture need to be returned to a "use stream" where nothing is wasted. The pattern of use and reuse in nature can continue to inspire our attitude toward the materials we take from the earth and mold to our convenience.

If we learn nature's lesson well we will find ways to meet the material needs of people without creating waste. Otherwise, we may be only a blip in the long record of history, where we turned all of earth's precious resources into garbage, and, instead of fragrant old growth forests we leave to future generations islands and mountains made of garbage, and barren hillsides where beautiful trees once grew. Instead of producing garbage, we need to reverse the process and follow the model of the earthworm, who ingests the earth's garbage in one end and produces fertile soil for new life from the other.

5.

The Ecology of
Making Things

American industry can begin to imitate nature through creating "closed loop" manufacturing processes. A closed loop process allows no waste and every by-product becomes the basis for a new product. As each industry with closed loop manufacturing finds its appropriate niche in the business ecosystem, all of industry can move toward being a climax community where no excess or waste is tolerated.

In the Chesapeake Bay estuary, a great blue heron with a speared fish on its beak glides through the marsh on the outer edge of the bay. A few feet away, in the shallow water, a canvasback duck dives to feed on eelgrass, ignoring the crabs scavenging nearby and the oysters clinging to the underwater rocks. In the deeper open waters of the bay anchovies and rockfish swim, freely eating plankton and each other. Above them ever-watchful gulls and terns watch for a silvery flash in the water below which could be their dinner. The profusion of life in the bay is the end result of a whole complex system of production by nature, which makes it one of the most abundant fisheries of North America.

In places like Chesapeake Bay, where life flourishes, nature manufactures—food and places to have and raise the next generation—the necessities of life with ease and grace, a sharp contrast to the inefficient way in which humans have created their industrial processes. Industry gathers raw materials from sources in nature, uses huge amounts of energy to create products out of these materials, and throws away any waste created in the process, often generating toxic by-products. The final products of most manufacturing processes are not designed with reuse in mind, so that after the original use is finished, the final product, just like the waste created during manufacturing, is thrown away. Obtaining the raw materials is usually done in a way that creates even more waste and pollution, such as the tailings left after a mining operation, or the eroded hillsides from the clear cutting of timber.

Nature's manufacturing plants, on the other hand, weave complex elements together with maximum productivity and minimal toxic by-products. In Chesapeake Bay, countless billions of microorganisms, mainly bacteria, live in the mud, feeding on dead fish and vegetation, providing carbon dioxide for the plankton and other aquatic plants which in turn are eaten by the fish and filter feeders. A balance is obtained as nutrients cycle from life form to life form. Filter feeders, like the oysters, clean the water and adjust the salinity by eating plankton and other microscopic plants and animal food floating in the bay. The marsh filters out contaminants from the water that comes from the surrounding waterways and creates habitats for birds who eat the fish and contribute their own waste back to the system. Energy efficiency is enhanced by the constant transformation of potential pollutants into useful material. Each niche is occupied, creating "jobs" for all kinds of organisms.

The elegant loops of nature's manufacturing processes—life to death and back to life again with no waste in the process, materials constantly being moved and transformed—serve as a model we should strive to copy. Currently this is happening in a movement called industrial ecology, whose goal is to encourage manufacturers to emulate nature's efficient and sustainable use of resources. The members of the movement advocate "closed loop" production in which the materials brought into the manufacturing system are used completely, with the most efficient use of energy in their production. This movement represents a transformation in the way we view our relationship with the materials we take from nature, from one of exploitation to one of cooperation and interrelatedness.

One of the foremost theorists of the practice of industrial ecology is Hardin B. C. Tibbs. His principles can be summarized by what happens in Chesapeake Bay: input and output must be in balance, and the efficiency of energy used must be maximized. As he outlines principles of what a system of industrial ecology would look like, he believes the transition from our current industrial process to industrial ecology will move in the following distinct stages: outright denial that any change in industrial process is needed; compliance with government regulations; and finally a full blooming of innovation and creativity as the new challenge of working within nature's limits is accepted and embraced.

When we look at American industry we find many examples of businesses and industries that act as though they are still in the denial stage, publicly professing that none of this attention to the environment is really necessary. But most of those who publicly denounce the need to consider the environment are working behind the scenes to understand what changes must be made in future products in order to have them ready to meet new laws and regulations. For example, all the major auto makers in the US are gearing up to produce electric vehicles to sell to California when the new emissions laws take effect in the late 1990s, while at the same time they continue to fight against regulations requiring increased efficiency and better mileage in current models of automobiles.

Whether expressed publicly or only being considered in the boardroom, a new awareness about their responsibility to the environment has companies everywhere looking for ways to solve environmental problems. Tibbs's next stage of industrial transformation, compliance with regulations, is beginning to transform the computer industry. Federal regulations on clean water and air date back to the late 1960s, but were updated and made more comprehensive and strict in the late 1970s and most recently in 1991. In Silicon Valley, south of San Francisco, cities vied for the so called "clean" computer industry as new information technologies blossomed in the 1970s and 1980s.

The cities didn't know, however, that computer parts manufacturers, such as those who make silicon chips, used highly toxic solvents to clean the chips during manufacturing. The computer industry mushroomed and these solvents were then disposed of as waste that frequently entered the groundwater or was discharged into San Francisco Bay. The problem of toxic pollution came to light many years after the cities believed they had solved their economic problems, and avoided environmental pollution, with a clean industry.

In response to changing regulations, in-house employee pressure, and a desire to present an image as socially responsible, an organization of computer manufacturers formed in Silicon Valley, dedicated to finding nontoxic chemicals to replace what the toxic solvents do. Apple, Hewlett-Packard, and all of the big chip makers, such as Intel, are involved in this effort. Some parts of the cleaning process that formerly used toxics are responding to such simple replacements as citrus juice derivatives, or even hot water.

New actions by the EPA aim to help companies past the second stage of industrial ecology, complying with regulations. By giving grants to research facilities to discover new nontoxic alternatives in manufacturing processes, the EPA will help companies find alternatives to substances which pollute the air and water. Some of their successful grant recipients so far have been John Frost, Purdue University, for cornstarch with bacteria to replace some uses for benzene, a known carcinogen; Gary Epling, University of Connecticut, for the use of food dyes and a sunlamp to replace several toxic chemicals; and James Tanks, Virginia Polytechnic Institute, for a new form of carbon dioxide to replace benzene and carbon tetrachloride. And although not one of Mr. Tibbs's stages, flexibility is a key attitude to help make industrial ecology a reality, where regulators help find solutions as well as enforce laws.

Some companies are at the next stage in the progression toward industrial ecology: cleaning up their wastes, recycling unused materials, and taking some products back for partial recycling. Herman Miller, a manufacturer of high-quality office furniture, is a well-established business that is beginning to see the profit of following nature's examples. In 1990 Herman Miller banned the use of rosewood and mahogany in its furniture line as a result of looking at the environmental impact from the logging of these woods in the rain forest. But that was just the start.

The company began to look for and use recycled materials, recycle its own products, and produce a long-term plan for eliminating toxic products from the manufacturing process. In the office and factory they purchased five thousand ceramic mugs and banned the use of Styrofoam cups. Then they began to sell eight hundred thousand pounds of scrap fabric a year that had formerly gone to the dump. The scrap now goes to a plant in North Carolina where it is used as insulation for car roof linings and dashboards. The company even buys back their "Action Office" furniture line, reconditions it, and resells it.

Herman Miller, now thoroughly committed to making changes ahead of coming legal restrictions, has not always made money on these early decisions, but they realize that this is a time of transition. But some companies feel confident in their ability to make profits from imitating the ways of nature. Three M Corporation is one of them. They project if they can achieve a 1995 goal of reducing air and water emissions by 90 percent and solid waste by 50 percent, it cannot only make a greater profit but it can then cut the cost of products sold to the public by 10 percent.

Finally, there are a few companies creatively trying to adopt industrial ecology, through actual closed loop production. Deja, Inc. is a new company founded specifically to develop environmentally responsible products. Its first product is a line of shoes that are made with significant amounts of the waste products of other manufacturers. Wet-suit trimmings, seat cushion trimmings, and the trimmings from disposable diapers are all used in the shoe. The company sends the shoe out in a box of recycled cardboard with drawings of nature scenes in the inside which is designed to be turned inside out and used as a gift box. And the company will take the shoes back after they have been worn out and recycle them. Although this company is small it can serve as inspiration to entrepreneurs, who make up the fastest growing segment of the economy. Maybe even the large companies will learn from their successes.

Another business, Anita Roddick's Body Shops, founded on being ecologically responsible and aiming toward closed loop production, started small and is now a worldwide chain. The Body Shop sells soap, shampoos, lotions, cosmetics, and bath accessories like candles and bath mats. Beginning in London fifteen years ago, Roddick decided to offer a line of personal care products which emphasized natural ingredients and recyclable bottles. She has been a leader in searching for new ingredients for her potions in the rain forests and jungles of the world. Not only do Body Shops feature products made from natural ingredients obtained in a sustainable fashion, but they also feature a constantly changing supply of educational material on the environment and social issues.

Ms. Roddick buys many raw ingredients from indigenous people in the rain forest, encouraging them to create their own sustainable economy and save the trees. For example, many Body Shop products use brazil nut oil, which comes from harvesters who also now process the oil in a factory near their homes which Ms. Roddick's company helped to set up. Similarly, The Body Shop obtains blue corn from Native American growers in the Southwest for use in a variety of skin care products. And, for anyone who doubts that a true commitment to the environment is profitable, there are now twelve hundred Body Shop outlets worldwide.

Although there are companies pioneering changes in their use of the earth's resources such as Deja, Inc. and Body Shop, it is clear that industry in general is responsible for the major share of depletion of resources and pollution of the environment. They do the mining, clear cutting, disposal of toxic wastes, carbon dioxide emissions, and release of the soup of chemicals in the air that creates acid rain. But we buy the products made by these environmental monsters, so our opportunities and responsibilities to influence the course industry takes are enormous.

We can influence industry's behavior by changing what we buy, by electing officials who will pass laws to regulate industry, and by investing our money in those industries that are environmentally trustworthy. As American industries make daily decisions about how they will operate they are, of course, trying to make a profit. They need to learn that they can make a profit and be environmentally responsible. In fact it is becoming increasingly clear that environmental responsibility is better for the profitability of business.

If instead of paying to have waste products taken away, industries are able to sell them, it is more profitable. And as laws are passed to prohibit industry polluting the shared domain of air and water, it becomes more profitable to clean up wastes rather than pay the fines and face public outcries that could reduce sales. Finally, industry has lost enough lawsuits for injury due to their use of toxic chemicals or their production of unsafe products, that, again, it is more profitable to start to clean up their act and try and avoid legal action. Industry is embedded in society just as individual plants are embedded in nature's ecosystem; neither can exist without the mutual support and cooperation of others in its system.

A system of industry that offers closed loop recycling, waste-free packaging and products, and a truly synergetic system for industries to interact and exchange wastes and products is still in the future. But we are moving in that direction. Small entrepreneurs and big businesses are actively searching for ways to be more environmentally responsible and make a profit doing so. The pattern of nature is gradually being transferred to the industrial process.

Americans look to business initiative and entrepreneurial creativity to promote change, but in other parts of the world the next steps toward industrial ecology are being fostered by government. In Japan the New Earth 21 Project, aimed at the threat of global warming, was initiated by the government. The goal is to develop technologies to reduce the production of carbon dioxide and other greenhouse gases. Tax breaks and low-interest loans are available for projects that meet the requirements.

In Britain the government is giving money to support innovations in environmentally friendly industries. In Germany the government is investing in solar, wind, and wave energy research, and German commercial banks give low-interest loans to projects that protect the environment.

These governments believe that the next wave of industrial change will be toward industrial ecology, and they are trying to give their countries the advantage in the global marketplace. The current global market for environmental friendly products is now estimated at $200 billion/year, and that figure will grow exponentially in the next decade.

So for both our environmental and our financial health we are fortunate to have a new administration that is actively promoting a shift in our industrial processes. In June of 1993 the Clinton administration announced the appointment of the President's Council on Sustainable Development. This twenty-five-member group is charged with sponsoring demonstration projects by public and private sectors, educating the public on the need for sustainable development, and establishing links with other countries for research and development purposes. Finally, after twelve years of standing on the sidelines as the rest of the world developed policies and industries to move toward a more sustainable relationship with nature, America is taking its place in this important work.

To achieve ecological balance in our industrial processes we need to recognize that every part of the industrial process is important. As we learn to substitute nontoxic chemicals, to find markets for byproducts, and to make products that are recyclable, we come closer to being in harmony with nature's designs.

As consumers we have great power to determine the course that industry takes. If we insist on goods that are recyclable and made by closed loop, nontoxic processes, someone will fill that niche.

6.

Leaves Never Hide
from the Sun

Efficiency is a basic concept in all of nature's activities. It is always present in the cooperative understructure of nature where life never uses any more energy than necessary to get the job done. In our age of constantly increasing pressure, we need to learn nature's lessons and conserve the energy we spend in our daily activities, seeing that a simpler life is more rewarding emotionally and spiritually.

Nature is not so much red of tooth and claw as it is full of naps and rest periods, especially for the longer-lived animals. Familiar pictures of lions lazing in the sun come to mind. Adult animals do what they need to sustain life. They do not, for example, waste energy stalking prey if they are not hungry. Domesticated dogs and cats with reliable food supplies spend much of their time sleeping and resting. In general, the longer-lived the organism the more measured its use of energy. Giant tortoises lumber slowly through their hundred-plus years.

Nature misses few tricks when it comes to finding ways to use available resources with the least energy possible. After millions of years of evolution, plants and animals function in the most energy-conserving ways; we humans, sabotaged by our energy-wasteful habits, have much to learn from nature's resourcefulness.

In the plant kingdom, the constant evolutionary quest to use energy more efficiently can be seen in the patterns of leaves on trees. Leaves are always spaced so that all receive some sunshine during the day. It would be a waste of the tree's energy to produce a leaf that would not have the opportunity to photosynthesize. Other plants save energy in different ways. Desert plants turn their leaves so the sun strikes the edge, minimizing water loss. Succulents in the desert develop thick skins to keep the moisture in their leaves from evaporating, thus saving their lives between the infrequent rains.

Even events that may seem to us destructive and wasteful provide opportunities for nature to be efficient. When the spring rains pelt the landscape so hard that runoff raises the creeks and rivers to flood levels, trees are pulled down and wildlife habitats are disrupted. But new growth quickly appears in the sunlight let in by the removal of the trees, and silt from the river bottom is deposited on the banks of the river, giving new fertility to the soil. Similarly, fire, as it rages through a forest or a prairie, consumes the dead vegetation where there was not enough moisture to cause its decay, and returns the elements trapped in the dead leaves and grasses to a usable state, again enriching the fertility of the soil.

In another careful use of energy, the plant kingdom has evolved to respond with great sensitivity to seasonal climate variations. Plants slow down their processes in a northern winter and open their tender buds at just the right time to take full advantage of the warmth of spring. In areas of continuous ice and snow, hearty lichens have evolved to take advantage of the brief periods of sunshine. And in the tropics when rain is available, growth continues all year around.

Nature doesn't try to transport water to grow roses in the Sahara or create greenhouses to produce orchids in the Arctic. The efficiency of nature is direct and straightforward. Water follows the line of least resistance and is stored in the lowest part of its path. Plants gather the nutrients from the soil around themselves, and without human intervention, only the plants that are suited to a particular soil type will grow in that soil.

In natural succession from pioneer to climax community each stage of succession fits exactly the energy available. Pioneer plants do not seek lushness, but sink their roots deep into the barren soil for the available water supply. Herbs and bushes that follow pioneer plants also are deep-rooted but their growth is more dependent on the nutrients fixed in the soil by the pioneer plants. And finally, trees grow tall, deep-rooted, and thick-leafed on the soil created by the pioneer plants, herbs, and bushes.

Nature is not capricious in the use of energy. And nature never uses extraordinary resources to pioneer the development of a new plant or to keep an outmoded species alive. The trees don't try to sprout their leaves when it is too cold and use excessive amounts of their energy reserves to keep the leaves alive in inclement weather. Roots don't grow so far from the tree that an insignificant amount of water is returned for their effort. Nature always knows when the effort is enough to justify the means it has used.

Humans, however, both in our creation of technology and in our current cultural behavior patterns, have not learned from nature. Not only does our technological structure waste nature's precious resources, but we seem unable to create work and leisure patterns to make rational use of human energy.

Our industrial system of providing energy was built on an abundance of raw material that we have used with reckless abandon. Our model of industrial technology evolved from the idea that the raw materials of nature, which supplied energy, were there for the taking and were essentially free to those who could wrest them from the ground. Coal, and then oil, have been relatively easy to obtain with cheap labor, and with no cost to the producers for the environmental damage they created. It didn't matter how quickly the machines and engines consumed raw materials.

Monuments to industrial inefficiency are scattered throughout the Midwest, sometimes called the rust belt, in the skeletons of huge plants where our industrial processing once occurred. And the land is littered with abandoned mines and quarries where the raw materials of industrial society were obtained. The process still goes on as we strip the tropics of their forests searching for minerals, or to raise cattle, and selling the trees cheap, when we bother to cut them instead of burning them on the spot.

But after several hundred years of wasteful use of our energy resources, we are now learning that we can design almost any energy use we have to be more efficient. Light bulbs, refrigerators, water heaters, and other household items are in transition. The thoughtful design of these items toward greater efficiency is beginning to happen in this country. Instead of more BTU's the new fuel of the twentieth century will be brain cells. Everything we will do needs to be smarter, with more careful thought and planning.

Part of the first wave of redesign for energy efficiency has been the lowly light bulb. In this country a large percentage of electrical use goes to lighting our buildings. Traditionally we have used fluorescent lights only in the office or factory. A revolution in the design of fluorescent lights has eliminated the buzzing noise, enlarged their color spectrum, and made them infinitely more efficient for both home and office. The new fluorescent bulbs use about one-quarter the energy of traditional incandescent lights and the bulbs themselves last ten times longer. A typical household can save 75 percent a month on the portion of their electric bill that reflects lighting. The drawback is the up-front cost of the new light bulbs, which cost ten times as much as standard bulbs. It is an investment that will save the user money, but requires that the purchaser see a light bulb as an investment rather than a simple purchase.

The person most responsible for the new look at light bulbs is Amory Lovins. In 1976 Lovins, energy guru and co-founder of the Rocky Mountain Institute, began looking at the least painful ways to introduce efficiency into the US system.

Lovins wrote a book in the 1970s called *The Soft Path*, which showed the economic and environmental benefits of using conservation. He predicted that with the use of energy conservation energy use would go down even as production went up. Government and industry predicted that energy demand in 1990 would be almost double of what it was in 1975. Actually energy increase from 1975 to 1990 was only 10 percent. Utility companies and other energy producers in this country laughed when his ideas were first introduced. But fifteen years later, Lovins's predictions have come true and he now consults with the utility companies and businesses who said energy conservation would never work.

Lovins and other experts on energy conservation talk about *"mining"* conservation, where mining means that more energy is obtained from each new level of efficiency of the design for a product. So, as the average miles per gallon on cars gets better, we can drive the same number of miles with less fuel. The same is true of efficiency in any appliance with more efficient design, less energy used even with the same level of performance.

Major utility companies are now taking conservation as a dominant theme in their long-term planning. Pacific Gas and Electric Company in Northern California, the largest utility company in the country, has created a plan in which conservation allows the company to postpone and possibly even cancel the building of new power plants. PG&E is now investing in more efficient lighting, heating, and cooling systems to wring more BTU's out of the oil and natural gas than they now burn.

PG&E, headquartered in San Francisco, is one of the most innovative utilities in beginning a transition to sustainable practices. But it wasn't always that way. During the 1980s PG&E was building a nuclear power plant and was in constant legal disputes with environmental groups. Lots of time, energy, and money were spent on lawyers, and for ten years PG&E was seen as the great spoiler of the environment. During the 1980s PG&E also lost a number of its large industrial customers, such as Chevron. These large companies decided to generate their own electricity cheaper rather than pay PG&E rates. As the end of the 1980s approached, PG&E felt very much in need of self-evaluation.

As an institution they wanted to be part of the good in the community, to be seen as solving problems instead of creating them. PG&E was in the enviable position of having enough resources to produce the electricity needed for the near future. They could afford to look ahead and try to understand what they should be doing in twenty years. So in a remarkable turnaround, they began to develop alliances with environmental interests rather than fighting them in court.

Today PG&E is selling conservation as well as electricity. The basic idea is that by selling efficiency to customers, a utility company reduces its need to build costly new power plants and can therefore make money on selling less electricity by eliminating the construction costs of new plants from their budgets. And if they sell enough conservation and efficiency, they can gradually eliminate the least cost effective ways they have for producing electricity and still make a better profit. So PG&E followed nature's model and began to sell efficiency.

There is currently a lot of activity going on to inspire both public and private companies to design and produce energy efficient uses. PG&E and a group of other utility companies offered a prize of $40 million to the manufacturer who could produce the most energy efficient refrigerator. And the Department of Energy is offering $432 million for "market mobilization" to encourage states, utility companies, and industry to proceed with research and development on conservation.

Another large energy use is in the heating and cooling of buildings. In this area two of the choices we have in energy conservation as we move toward a sustainable society are 1) to develop efficient means of cooling and heating our buildings, or 2) to learn to live more in the rhythm of temperature changes. It is hard to imagine our society without the heating and air conditioning which we use so lavishly in our homes and workplaces. But, with properly designed heating and cooling systems we can be more flexible in heating and cooling only the spaces we use and reduce our energy use.

Efficiency in using nature's resources is not a new concept. People have been devising ingenious solutions to make more efficient use of their energy supplies for millennia. In China heavily padded clothing in the winter instead of heating buildings is an old solution to keeping warm. Another variation on the theme of heating only what was needed was in Japan where a hibachi was placed under the dining table and a large heavy tablecloth was tucked around those seated at the table to keep the heat in. And the Eskimos built small, tightly insulated shelters that were heated by oil lamps and a critical mass of human bodies.

Many nontropical indigenous people have devised ways of getting clean and warm at the same time with a minimum use of resources, an efficient use of both their resources and their time. Northern European peoples used saunas or sweat baths to both warm people and clean them. With a small fire, a small shelter filled with people can get quite hot. Rocks are heated in the fire and hot water is poured on the rocks to create steam. The same thing is done by Native Americans who call it a sweat lodge and get even greater efficiency, because not only does it warm and clean them, but it is a source of spiritual nourishment. The Japanese had large public baths to warm and relax people before going to bed at night.

An equal number of devices have been created by people trying to keep cool. Desert regions have many variations on the air scoop, a tall chimney structure which sucks the cooler air from the ground into the building and sends the hot air up the chimney. The hand-held fan is a universal device. The wealthy and important generally have someone else wave the fan, usually a large, ornate one. People in the tropics wear little clothing if they live primarily in the jungle shade, or loose airy robes if they are primarily in the desert sun. The most universal adaptation to the heat is the afternoon nap. All non-air-conditioned tropical cultures come to a halt when the sun is at its hottest in midafternoon. People sleep until the worst of the heat is past and then work again as the evening begins to cool down.

Will we ultimately choose to return to the slower lifestyle of the past more in tune with nature? Before air conditioning, lazy summer days meant afternoon naps and reduced work loads. We remember visiting a tiny town in the tobacco growing area of North Carolina for a summer wedding. As part of the wedding party, four of us arrived early for the rehearsal. After the rehearsal we were sent to an elegant columned house not far from where the bride lived. The gracious lady who met us at the door ushered us upstairs to the bedrooms and installed the ladies in one room and the gentlemen in another. In the sweltering afternoon, nap time was for resting, and nothing else, as the deliberate separation of married couples underlined.

Now, we not only work during the summer's wilting heat, burning large quantities of fossil fuel to keep ourselves cool, but we work all night putting another kind of stress on the human body, and using more fuel. Night workers, who frequently alternate between day and evening shifts, are even less efficient as they exist in a state of constant jet lag.

The reality of the 1990s is that even people with regular day jobs work harder and longer than at any time since social reformers began to pass laws to eliminate the "sweat shops" of the nineteenth century. The complaint of being overworked is heard from all segments of society (except the unemployed). Professionals are expected to do at least 50 percent more than their previous jobs due to downsizing of companies. Hourly workers are tapped for overtime to avoid putting on new staff and paying benefits to them. The syndrome of overwork is bad enough in this country, but in Japan the medical profession has identified a new cause of death of epidemic proportions: "karoshi," or dying from overwork.

The stress of an extra work load for most people is compounded by a long daily commute that now stretches up to three hours, each way, per day, in the countryside of major metropolitan areas, as people search for an affordable place to live their image of the American Dream. Our current patterns of work and life consume large quantities of nature's raw materials and erode our physical and mental health.

As we saw in Part I of the book, a sustainable society of hunter/gatherers only spends a few hours every day meeting their basic needs of food, clothing, and shelter. They have time to play, tell stories, and engage in religious rituals. Our overworked twentieth century psyche yearns for time to be human. How efficient can it be to make some people work past their endurance while other people have no work at all?

It is difficult to imagine Americans deliberately choosing to slow down their lifestyle to use less energy. But it is closer to nature's way—a way of greater harmony with the seasons and varying climates of the earth. Perhaps someday our descendants will look back at us and wonder how their ancestors survived as they turned away from the rhythms of nature to embrace twenty-four-hour-a-day shopping and climate control that wastes enormous amounts of energy and makes all times and places feel the same.

Nature remains the ultimate master at efficient use of resources: not a leaf on a tree or a drop of water is wasted or used to less than its full potential. By contrast, humans have squandered the bounty of nature as they have discovered how to use nature's resources. Now we find that many natural resources are not as plentiful as they were for our parents and grandparents. We must learn new ways, such as using energy-efficient light bulbs and appliances. And we must search for better ways to use human resources, allowing all to be productive without overwork or unemployment. We can learn new ways, for that is our talent—to invent, to create, to understand the world around us, and find the best ways to live in it.

7.

The Wisdom of Motion

Transportation is an area where an attitude change is necessary. Nature simply doesn't commute to work. By following nature's examples we can find ways to create more rational ways to determine where we live and work. New cooperative community patterns can help us make a transition to a society where movement is based on principles of sustainability.

Nature's transportation schemes have a beauty borne of necessity. The grace of maple seeds as they glide through the air, whirling like tiny helicopters, is a result of the need to get far enough away from the parent plant to find an open space in which to grow. The sleek beauty of the cheetah or leopard is the result of bones and muscles honed to minimum mass and maximum strength to create the burst of speed necessary for the animal to catch its dinner. And, although the flower is rooted in the earth, its color and fragrance encourage insects to transport the flower's pollen far and wide.

Most plants and animals evolve to move within a certain set of environmental conditions. Reindeer thrive in harsh lands by slowly migrating within the zones of the northern temperate forest. Following an annual cycle of available food and water, even birds choose routes where they can stop for nesting locations which afford maximum safety and ease for their species. Transportation in nature is based on getting what you need from as nearby as possible, using as little energy as possible to get it.

A good example of how energy is conserved when movement is necessary in nature is the way plants are able to obtain water. Even though plants can't walk, they move enormous quantities of water through their systems daily. The process of getting water from the ground to the leaves of trees seems to work on the suction or vacuum principle: pressure creates a vacuum that pulls the liquid up. Without expending any energy the tree simply stands and takes advantage of physics at work.

Plants also do very little work when it comes to spreading seeds, using strategies of great variety. Many seeds are scattered by birds and animals. The animals eat the fruit of plants, and the seeds become part of their eliminated waste, which can be very far from its source. Birds carry seeds on their feet and deposit them in far distant locations. Many seeds are shaped so that they are carried on the wind or stick to the coats of passing animals. In none of these instances does the parent plant do any work, except the initial design of the seed delivery package, where they capitalize on the natural conditions available.

Seeds are an excellent example of how the initial design of a system is the most important component, taking into account all stages of what the seed must do. The plant must only expend so much energy to produce it in the first place. Then, the seed must also be protected from the elements and predators, either with a protective shell or by being so numerous that some will survive. Finally, the seed must have the possibility of reaching a place where it can put down roots and sprout.

As we turn to the human uses of transportation, we need to examine the objects and systems we have created, such as automobiles and the daily commute, and the motivations that humans have for transportation and travel. What can we learn from nature that will aid our transportation needs? Can we discover the necessities of how, where, what, and why we must move in such a way to give our efforts the ease, beauty, and economy of nature?

If we look at our own past we find that movement and travel were easy for humans when we were still hunter/gatherers—we walked. And, since we ate and made our homes and clothing out of what was nearby, we didn't need any sophisticated methods of transportation. Many hunter/gatherer groups lived seminomadic lives depending on the availability of food. Some had permanent summer and winter camps, and traveled between the two according to the season. Others had one large permanent camp they would use when food was plentiful, but would disperse into smaller groups and fan out across the countryside when food was scarce. These transportation patterns served to ensure that they could secure food from their surroundings in such a way that food would continue to be available year after year.

But movement and travel for hunter/gatherers was not just for finding food; a large number of hunter/gatherer groups also went on long journeys to visit sacred places. The Australian aborigines still go on a "walkabout" to visit sacred caves, mountains, and springs. Native Americans similarly journey to sacred sites such as Bear Butte in the Black Hills of South Dakota and Mt. Shasta. They were driven by inner spiritual urgings to visit these places which focused and stabilized their tribal communities.

Wherever they went, hunter/gatherers traveled light. Their few possessions were easily carried, they lived off the land, and their home and clothing were easily duplicated from the plants and animals they encountered.

Much later, agriculture and then industrial civilization bound humans firmly in one spot. Movement of goods increased steadily as we first harnessed animal power and then machine power to help us move products to be sold and traded. But until the advent of the industrial age people themselves lived close to where they worked, and for the most part they lived in houses made out of local materials, wore clothes produced nearby, and ate the food of their region, in a pattern similar to that of the hunter/gatherers. This pattern of local use was destroyed by large-scale trade made possible first by the steam engine and the railway system, and then by the internal combustion engine and the automobile.

Now, in total contrast to nature's schemes and our own history, human transportation is one of our most energy-intensive and polluting activities. And the transportation patterns we have already established may be among the most difficult to change as we move toward sustainability. Humans love mobility. Riding in fast cars, airplanes, even chugging along on a train are all pleasurable activities to many people where we not only travel for work and fun, but continue our ancestor's pilgrimages to far places. The kind of transportation we have developed, especially in America and Western Europe, allows us great freedom of movement where traveling to distant places is a prime recreational activity in which people take great pleasure.

Our designs for transportation systems have been based on raw power use, rather than a clever imitation of nature. We move things around in frenzied patterns, wasting time, energy, and resources. Our homes and work places are frequently far apart, making daily trips, frequently as a single person in a private automobile, a necessity. The elements used in our food, clothing, and housing are gathered from all over the world with little thought for the energy spent in transporting these items.

The environment has begun to feel the effects of our excessive movement. The oil necessary to fuel our activity must be recovered from the earth, and, since we use such large quantities of oil, we are willing to search for it anywhere, even if it means destroying sensitive ecosystems like the Alaskan tundra. The transportation of oil is an ongoing threat to the environment as evidenced in the frequent oil spills around the world. The daily use of so many cars and trucks has created a brown haze of chemicals hovering over many of our cities. And, finally, to make places for the cars, huge tracts of formerly productive farm land are covered with asphalt. We are obviously not copying nature's ways.

In order to imitate the wisdom of nature when it comes to transportation, we need to constantly ask two questions: Does something need to be moved? If so, what is the most energy-efficient way to move it?

One of our biggest transportation expenditures, and one that people are starting to question, is getting people to and from work every day. Millions of people spend hours in their cars each day commuting to and from work. The situation is beginning to be addressed by many companies, who are saying: let's take the information to the worker, rather than bring the worker to the information. With the advancement of modern communication systems, for example, there are many areas of information and data processing that do not require workers to be present at a particular place. That explains why, when you pay some of your monthly bills, you may be sending payments to small towns in the Midwest when the corporation's headquarters are located in large Eastern cities.

The companies save money by hiring people who work in isolated places where wages, rent, and the general cost of living are lower. These people never go into a central office. Some work out of their homes and some work in offices set up in these small towns. Citicorp has its credit card operations in South Dakota; Patagonia has its customer service staff in Montana. And Lusk, Wyoming, population one thousand, is installing fiber-optic cables to attract more of the same kind of business.

This telecommuting trend is not limited to data processors; managers, owners, and some highly trained technical personnel are also loading the office data on their laptops and heading for home. Money managers are leaving Wall Street and heading for the backcountry; with computer technology, trading can be done from anywhere. Although most of these people did not make their decision to work at home to reduce pollution and save resources, that is exactly what they are doing.

And, of course, the final category of telecommuting workers are all the many entrepreneurs starting their own businesses out of their homes. Downsizing of large corporations is the trend of the '90s, and many of these displaced workers are going to work for themselves, at home.

Beyond reducing transportation problems the advantages of telecommuting are easy to see: reduced costs in almost every category. Businesses save on office space and the need to provide parking for their employees. Employees save the commute time, cost of gasoline, and wear and tear on their cars and on their psyches.

There is, of course, the need most people have to be in direct contact with other people. There are plans in communities across America to open offices for telecommuters who want to avoid the commute but would like to be where other people are working. In these local offices, people of different professions working for different companies will share office space and still be a few minutes from their homes.

In 1993, twenty-six hundred of the eighty thousand people employed by the County of Los Angeles were telecommuting. The trend is bound to accelerate in the Los Angeles area, where even in the middle of the day congested freeways can bring traffic to a crawl, and people often commute fifty to seventy-five miles each way to work. The telecommuting office of the Los Angeles County government estimates that by the year 2000 as many as 25 percent of their employees may telecommute.

The possibilities are endless. The car pool might become the work pool using a spare room or basement that one person might have, leasing space in a neighborhood work center, or using vacant space in neighboring city and county government offices as the County of Los Angeles is doing. People who can arrange it might move to small towns where the quality of life is better and living costs are less. And the quality of life will improve even more as unnecessary trips are eliminated.

Another area where we need to ask the question "What needs to be moved where?" is the transportation system, which moves food, clothing, and everything else we buy and use. A concept that addresses our need to limit unnecessary movement of goods is bioregionalism, which stresses the need to develop ways of living more off the goods and services produced closer to home. A bioregion is defined by its physical characteristics such as, microclimates, watersheds, and vegetation rather than arbitrary political boundaries. According to this definition a large urban area and its surrounding suburbs fall into the same bioregion.

Food is one of the most important elements in the bioregional concept, where there is a vision of people eating more and more food that is grown locally. Now, we buy and eat food grown thousands of miles away, which is very energy inefficient. A small part of bioregional self-sufficiency is becoming closer to reality with the constantly increasing number of farmer's markets in the country. Not only do farmer's markets cut down on the distance goods are transported, but they provide people with fresher food, and in many cases eliminate a middleman, allowing small farming operations to be more profitable, and enabling more people to enter farming and support themselves.

After we identify those things that don't really need to be moved, then we can look at how to move the things we do need to move in ways that are compatible with nature's examples. Remember, it is the initial design that will save energy, especially if that design imitates nature's constant quest to do more with less.

The best form of transportation whenever possible is always by foot; it is not only free and without pollution, but it gives the human body badly needed exercise. Next, nonmotorized forms of travel, like bicycles, can carry both people and some of their possessions faster and farther than on foot, again provid-

ing valuable exercise. Wonderful, large tricycles have
been invented which carry riders unable to ride bicycles.

When the weather is bad or people have to go farther than walking or biking permit, some form of
public mass transit is better than a city full of private
cars. But all of these ways that people move themselves and their goods around are made possible, or
not, by the basic design of our cities and workplaces.

In Europe and Asia, and even early America, cities
evolved so that most daily needs could be met by
walking. Residential neighborhoods were intermin-
gled with shops and small manufacturing operations.
The shopkeeper almost always lived behind or over
the store; there was no commute and less crime. A
mix of services evolved to meet the needs of the community. You could get food, have your shoes re-
paired, spend a social evening in a bar or pub, and
go to church without leaving your extended neighborhood. And you could do all of these things on foot.
In Europe even some farmers lived this way, living
in the community and walking out to their farm plots
everyday.

The automobile changed our settlement patterns as
cities expanded in America after World War II. People
scattered in all directions. Los Angeles, Houston, and
Miami ate up the agricultural land that surrounded
them and made most residents subject to the commute.

Now we are questioning the wisdom of the patterns
spawned by the automobile, and planning in cities is
beginning to tip back toward the compact, full-service
neighborhood. If you live in New York, Boston, Chi-
cago, or San Francisco you may never have lost this
urban compactness. And college towns all over the
country have maintained this pattern. Architects and

planners have begun to talk seriously about "in fill," where housing and commercial mixes make existing communities denser to take advantage of existing transit systems. Finally, the most popular new developments like Seaview in Florida are planned to have the services of a small town.

In Europe, where the car is a more recent intrusion, some communities are banning automobile traffic in their old inner cities. Cities in Switzerland and Denmark allow only mass transit in the inner city. This is done to control pollution, noise, and traffic congestion. But such action will have the added benefit of restoring charm and grace to these old cities, the very qualities that attracted tourists in the past.

Europeans have had the advantage of clean, efficient transportation systems due, in part, to the layout of older towns built for foot travel. In all but the smallest villages, you can travel quickly and easily all over town on streetcars, buses, or subways, in comfort and without fear for your safety.

From the early nineteenth century until the 1940s large American cities were served by a combination of rail and bus service that allowed a great deal of mobility without the automobile. There is some evidence that automobile companies deliberately bought up and dismantled some public transit systems. Then the federal government began to subsidize highway construction, leaving public transportation, frequently subsidized in other parts of the world, to fend for itself. This pattern is now being changed. In many places, state and regional governments are beginning to consider, with the blessings of the federal government, the use of funds previously earmarked for highways for public transportation. However, several factors make increasing both the quantity and use of public transportation a slow process.

First, the sprawl created by automobile-centered development leaves people scattered, so that the density of development necessary to make mass transit work well is lacking. The issue of sprawl is beginning to be addressed in places like Seattle where a new comprehensive plan for the city is proposing the encouragement of new housing development as "in fill" in existing neighborhoods to begin to create the density needed for mass transit. The plan also calls for the supplementation of its current bus systems with small vans to increase the range and flexibility of the system.

Second, to attract more riders our public transit systems need to be cleaner and safer than they are now. New York subways, and late-night buses in most cities, are so crime-ridden that only the poorest and those looking to prey on others ride them. In American society, which is constantly bombarded by media images, the image of public transit is very negative. Movies and TV programs portray bus riders as either too young or too poor to drive, and they show seats filled with characters who look threatening or like they need a bath and psychiatrist. In short, the image projected is of a place most of us would seriously try to stay away from.

Public transit needs to be clean, comfortable, affordable, and convenient to get Americans out of their cars. Even then it will probably also take legislation to limit the number of cars allowed into urban centers, and financial incentives to make it clear how much more expensive it is to drive a private car as a daily commute to work.

Even with the creation of good mass rapid transit systems in America, it is doubtful that we will eliminate the automobile. Therefore, we need to redesign it. The quiet luxury of a modern automobile masks its lack of efficiency. The automobile is a box on wheels used to take people and goods from one place to another. But it has become much more. The automobile is really an object of fantasy and dreams. People drive around suburban neighborhoods in Range Rovers which were designed for off-road travel that would challenge a tank. Limousines become traveling meeting rooms, bedrooms, and places to party. Porsches and Lamborghinis are designed to go at speeds that few of their drivers ever have the chance to test. When we go into a new car showroom we are dazzled by the wonderful curves of the body, the brilliant paint jobs, and the gadgets inside the comfortable interior. We are usually buying a dream, not a car.

Europeans and Asians have not been as ensnared by this dream and have always known how to do more with less. They have had energy-efficient cars and appliances long before the environmental movement was popular. The tiny cars that race around the streets of Rome, Paris, or Tokyo have never been imported to America. A whole family would squeeze into a car half the size of an American subcompact and feel they were rich because they had a car at all.

American automobile companies have consistantly shied away from small cars, but because of new emission regulations, especially in California, automobile companies are beginning to design small electric cars. This does not eliminate our excessive use of nature's resources, but, especially if electric car batteries are recharged at solar cell stations, it cuts down on pollution and moves our use to a cleaner and renewable energy source. The new electric car in production by General Motors is a sporty, sleek model, designed to activate your dreams as much as stir your civic desire to reduce pollution.

A boxy golf cart that goes twenty miles per hour is

not likely to win as many admirers as a car that looks like it could win the Indy 500. But for many of our daily tasks a boxy golf cart is all that is needed. In retirement communities in Southern California and Florida, residents do their daily errands in electric golf carts. The carts are easy to use and relatively inexpensive.

If we got everyone to work at home or commute to work by foot, bicycle, or public transportation, there is still the long-distance travel we do for pleasure, or the shorter trips for recreation. And for both of those we travel primarily by automobile. We need to rethink what recreation means, and then we may develop new images of pleasure.

Although many wonderful, complex, and sophisticated schemes may come from tying recreation to energy conservation and the desire to imitate movement in nature, sometimes the simplest are the best. A favorite of our family has been to take "urban hikes." With walking shoes and appropriate clothing, we walk the three miles from our house to downtown for dinner or a movie, or even a weekend trip to the library. It's nicest in the fall when the leaves piled up on sidewalks are crunchy underfoot, or in the winter, when a cold wind makes the restaurant at the end of our walk very inviting.

In addition to our simple, everyday needs for recreation, we sometimes have deeper urges to travel great distances. And when we travel long distances for pleasure it can become more of a pilgrimage, like our hunter/gatherer ancestors, a trip where the end result is to refresh and restore the spirit. If we are conscious that the end result of a trip is spiritually restoring, we may be able to make the means of getting there more restoring to the environment. Maybe we can come to see using a car as a sacred act because of all the resources sacrificed to make the trip possible, and cherish the freedom and delight it gives us while doing everything we can to make our routine activities as energy-conserving as possible.

Americans have plenty of developing models such as telecommuting, mass transit, and new designs for vehicles to look at to show them how to live with greater energy efficiency and move toward sustainable patterns in transportation. But we have been lulled for decades by cheap gasoline, wide open spaces, and the dream of the American automobile.

Changing transportation patterns is one of our biggest challenges in moving toward a sustainable society. With transportation we live in the realm of dream and fantasy. How do we reconcile the American dream of a fast and luxurious car with the reality of environmental pollution and resource depletion? Maybe if we can see that the dream is about freedom and opportunity, not cars, we will find new ways to understand the yearnings.

Even if it takes us awhile to understand why we enjoy movement so much, isn't it more important to leave our children and grandchildren a world with the earth's plants and animals intact than to mine all of our resources today to satisfy a brief cruise down the highway? It is no longer a question of *can* we do something that strikes our fancy, but *should* we do it? In making such decisions we need to consider the inherent wisdom of nature and the success of her transportation design.

FAIRY TALES OF TRANSITION

In Part I of this book we explored the past, examining principles that showed how natural systems function. Then we looked at how human evolution and behavior were strongly grounded in these principles. In Part II we examined the present, investigating some of the inner workings of nature to find models for revising our current industrial way of life. Now, in Part III, we will imagine the future, by telling stories of how transitions to sustainability might happen.

Many books on the need to make changes in the environment paint realistic scenarios of the future, usually emphasizing what a city might look like or how a new governmental system might function. We take a different, more fantastical approach. By telling stories we can show not only what systems and technologies might change, but we can also show how these changes might actually occur by giving a picture of the people who could make them happen.

In these stories we will seek inspiration from what we have already learned in the book: how cooperation helps people find a productive niche, and allows all parts of society to work together toward climax communities; and how imitating nature in our technological and social development can lead the way to a sustainable society. The seeds for renewal, just as in nature, are already planted and wait for the climate to change enough for them to sprout.

In nature the potential for change is always present in the background, where dominant species thrive because the existing conditions are right for them, and in the shadows and on the sidelines, different species exist ready to blossom. In the Eastern hardwood forests when the rainfall is moderate, allowing for dry soil, an oak forest will develop. But if the climate patterns change, the rains become more frequent, and the soil remains wet, and the oak trees begin to wither. They are replaced by the once-shy maples that were growing sparsely in a place too dry for them. As the new climate stabilizes and the ground stays wet, the maple flourishes and the oaks disappear.

An analogy exists for human society, where

the dominant way of doing things tends to push ideas promoting change to the background, until the right time and the right people come together. In the following stories we'll try to show the right time and the right people coming together. The time is during the next ten years, and the people are those who have both common sense—an inner voice that prompts them to see that change is needed—and the strength of character to act. They understand that humans are only one part of the ecosystem, and that to create a sustainable society we must work with and learn from nature.

History has clearly shown that individuals who have courage, determination, and vision can make a difference, and that apathy and resignation only lead to more problems. We hope these stories can inspire all of us to act with resolution and courage, and to work for the changes needed to allow the human species to thrive.

8.

The Greening
of the American
Automobile Industry

*As one of our largest industrial forces, the auto industry
could be a key player in leading the other producers of
goods and materials toward industrial ecology. This story
shows how such a change might happen. In the future
someone like John Carmichael, the hero of this story,
could transform the automobile industry. John brings a
new system of values emphasizing cooperation, a new relationship to the environment where he exibits the values
found in a climax community, and a concern about the
planet and how it is affected by industry.*

In the early 1990s a new spirit swept through the
American auto industry. Each of the Big Three
auto manufacturers acquired innovative CEOs
who could see that change was necessary, and they
began to attempt to revamp the ailing auto industry.
Although they had begun to think seriously about the
environment, their steps in that direction were tenu-
ous. They still focused on luxury cars that made a
higher profit margin.

Two of the Big Three auto makers put electric engines in existing models that were too big to go far on the available batteries. Or they trotted out designer prototypes that were years away from production when they wanted to show their environmental consciousness.

But in the workrooms and design workshops of the third auto maker a fire began to burn. Some of the engineers there thought they could produce a great new car. They dreamed of magnetic-levitation where the car hovered above the ground, hydrogen fuel cells, and cars with the total skin covered with solar cells. The cost analysis team began to dream of making something that would compete on the world market in terms of cost, and the few engineers interested in recycling began to dream of cars where all parts could be reused again. In 1995 this company got a new president who believed in dreams.

The new president was John Carmichael. He had been a dreamy child, full of fantasy and make-believe. But unlike many dreamy children, he enjoyed the outdoors. He had a great love of nature, and was fascinated by the mechanics of how plants grew. He also was aware at an early age of the environmental problems produced by industry, since in his hometown the snow would sometimes be pink or red with pollution from the nearby steel plant.

In his teen years he had begun to translate his fantasy into drawing and design. He was one of those rare individuals who was good at both science and art and loved the messy intricacy of biology as much as the precision of physics. Growing up in a suburb of Detroit made his desire to design cars as normal as making movies is to the children of Hollywood moguls.

In his late teens after he found he could use both his love of science and art in designing cars, John ex-

ploded in a burst of energy that never stopped. But
he remained secretive; a childhood of dreams and fantasies that other people laughed at had made him that
way.

He studied engineering in college and won numerous prizes for his design work even while in school.
When he graduated he was the first of his class to be
hired by one of the Big Three auto makers. Then he
ran headlong into the bureaucratic morass that helped
limit the design decisions in the automobile industry.
His ideas on styling were dramatic and saleable, but
he couldn't get anyone to listen to his ideas on materials use and engine modifications for better gas
mileage. He reverted to his childhood pattern and
kept to himself the ideas he thought would be rejected.

After a few years it became clear to John that more
decisions were made by managers than engineers, so
he went back to school to get a degree in business
administration.

John had grown up when the American car was
king of the road. He had lived in inward shame when
the Big Three didn't seem to know how to compete
with first the Germans, then the Japanese. John knew
what to do, but he had no power. He set his mind to
attaining that power.

John moved smoothly from middle to upper management. He was bright, had good ideas, and was
never too radical in his proposals. He had a knack of
finding savings and ways to cut costs without step-
ping on too many toes. His suggestions got accepted,
and he rose quickly. So in 1995 when a new president
was being looked for, his name came up.

This was what John had waited for all his life. He
had gotten this far by playing a waiting game, making
cautious decisions about how to do his best without

going too far in his suggestions for change. He decided it was time to change his strategy. The board would hire him and the board could fire him. He let them know his real plans.

John laid out a program for change and a vision for the future. The board initially wasn't sure what to do with his proposal. They had selected John as a clever but conservative cost cutter, and suddenly they had a radical on their hands. If John's proposal had come from anyone else, they would never have even read past the first page. But if John thought this was the way to go, maybe he knew something that they didn't. He was elected by a one-vote majority, and he went to work.

John Carmichael finally had the power he had worked for all of his life. He proposed several new directions. First, he proposed serious cost reduction in all departments, especially the elimination of unnecessary middle management and the containment of executive salaries. Second, he proposed the reorganization of all departments to focus on specifically stated goals. Third, he proposed strategies to bring design decisions to production more quickly. And, finally he proposed that the company adopt a true dedication to produce the best, cheapest, safest, most environmentally responsible transportation possible.

This strategy produced a revolt among shareholders at the next meeting, but John was able to convince them that the future of the industrial world was in creating long-range, ecologically responsible products. The day after his publicized speech to the shareholders meeting, there was a rush to buy stock in the company by millions of individuals and mutual funds who had made a commitment to ecologically responsible investing. The stock increased in value by 50 percent overnight, and the board of directors breathed a sigh of relief. John would have free reign, as long as they could see profits coming in a reasonable time and the value of the stock stayed up.

The company began to phase out any gas guzzlers still in their line. They were laughed at, but because of the shift in stock ownership were able to keep going. They downsized and diversified. They formed quality work groups. And they never forgot the goal of making ecologically sound transportation.

They used the engineers who were already planning for the future and created a department jokingly called the "Dream Committee" whose job was to come up with the most idealistic visions of future transportation, and then pass the ideas to research and development for a sifting through and shakedown of the most practical aspects of those ideas.

The Dream Committee envisioned a car that was totally recyclable, run by electric power, with solar arrays on the body to recharge the batteries whenever sun was available. They also envisioned selling each car with a solar-powered recharging station and an extra set of batteries so that there was never a drain on already existing electric utilities. The batteries were designed so that they could easily be removed and replaced as better battery technology was developed. Using 1995 battery technology, a car could go two hundred miles on a charge, and some big breakthroughs were expected to give similar batteries a four-hundred-mile range. On the horizon were batteries powered by replaceable fuel cells. Then the fuel cells could be sold just like gasoline anywhere in the country.

This dream car would be assembled out of interlocking and bolted pieces so that it could be taken apart back to its original construction. It was a mix of materials, but each material retained its integrity and therefore could be reused. Alloys and composite materials were designed to be reused. Recycled and recyclable plastic were developed that could be used in the car body. The plastic was a strong and resilient material that would allow cars in a collision to bump off each other like giant bumper cars or rubber balls.

The car was designed so that major maintenance and stress points were easily visible and could be checked by the car owner. Electric motors would need some cleaning and maintenance on brushes and rotors, but, of course, there would be no radiator, no oil changes and no points, plugs, carburetor, or fuel pump to maintain.

The Dream Committee knew that the most practical car would sell better if it had up-to-the-minute styling. So their practical, recyclable car was beautiful and aerodynamic.

The research and development division began to look at ways to buy recycled material for use in production and to devise plans for collecting and recycling the cars once they were through their useful lives. Automobile salvage yards are huge wastelands of wrecked, rusted, and worn-out cars, and have been part of the American landscape since there were enough cars to fill them. But what if the car company bought the car back at the end of its useful life? It could design how each part of the car would ultimately be recycled.

John Carmichael authorized the production of the dream car. Before John it had been a long time since a president of an automobile company had rolled up his sleeves and spent time on the production floors. But John had the background to understand what was going on and the desire to inspire and lead. He met weekly with the top designers and engineers, cheering them on and adding his own ideas.

Once the recyclable car was finished, it had its true test: Would the public buy it? They started with three models—a two-seater with barely enough room for a few bags of groceries, a hatchback that looked like the tiniest van in the world, and a sedan that would carry four people and had enough cargo space for groceries or luggage. Both as a marketing tactic and as a way to show the difference of the car, a whole new kind of showroom was designed. There were big displays showing the steps in manufacturing. The displays also showed what the car had been made from and how it would be recycled. The electricity in the showrooms was powered by solar energy and a large solar charging station charged all the batteries for the cars that were on sale. Salespeople were trained to answer questions and write up contracts, but never to pressure anyone in any way. The bottom-line prices were listed on the cars. All options were included, including a sound system. The first showroom was opened in 1998 and by 2000 the recyclable car was the hottest car on the market.

The simplicity of the design helped keep costs down; it also helped that electric motors are inherently simpler than internal combustion engines. And, the designers were fortunate that breakthroughs in battery technology brought down the price of the batteries, one of the most expensive components in an electric car.

By 2000 a new easily charged battery with a range of six hundred miles was available. Because it was electric, gas stations for this car were not needed. Because of the plastic reboundable shell, body work became rare.

The employees of John's company began to develop a great pride in their craftsmanship. Feeling that they were doing the right thing for the environment rubbed off onto all their other endeavors. The people on the line became very stringent about the bolts being tight enough, and defects being observed and eliminated. They had already begun to offer ways to improve their work areas and processes, but now they began to make suggestions about how the designers might change details to make assembly easier and the final product better. A whole new level of pride was evident as feelings of involvement and empowerment spread throughout the company.

Because the cars were compact in size, they were suitable for export to other countries. Sometime between 2000 and 2005 John's company became the first American manufacturer to export more cars to Europe and Japan than these countries were exporting to America. The Japanese were shocked. They had tried hard to keep the recyclable car out of the country, but consumer demand was so strong that the cars kept finding their way in. The Japanese finally gave in to their own public and allowed the cars to be imported in large numbers. In Europe the cars had captured the public's imagination in ways that had never happened before and sales were booming.

The export phenomena raised morale and dedication in the company even more. Finally, after thirty years of being beaten at their own game, an American car company was back in the world market.

John's company went on to examine what transportation meant in all sections of the economy and began to redefine itself as a company that produced transportation rather than cars. The company came up with many ecologically sound solutions to transportation problems. It used its design skills to serve as consultant to companies that manufactured streetcars, buses, and trains, and in addition to making them more energy efficient, the company helped to change both the look and feel of these weary servants of mass transit to make them more inviting to the public. By the time John retired the company was researching a magnetic levitation system that would reduce the costs and impact of road surfaces. It had become one of the most powerful corporations in America. John, after he retired from the company, went to teach at the Harvard Business School. He said that if he could change attitudes in the business world, half the work of transition to a sustainable world would be done.

9.

Growing Food
in the City

No self-respecting niche in nature would depend on resources as potentially unreliable as our modern farming
network. Grocery stores are the only storage that most
communities have. Natural or human-made disasters
could cut off that supply, causing social disruption and
starvation, especially in large urban areas like New York
or Los Angeles where inner-city decay has already taken a
great toll on both people and the environment.

The next story, presented in two parts, brings the vision of the book full circle to show how providing food,
restoring the environment, revitalizing a decayed urban
environment, and creating new opportunities for employment (or modern human niches) can all be part of the
same process. And, in fact, only a process this organic,
which weaves together many factors the way the ecology
of a natural system does, can be successful.

Both parts of the story focus on growing food in places
that have been damaged socially as well as environmentally, and in keeping with our earlier theme of cooperation and symbiosis we see that farming creates an energy
and enthusiasm that spawn new industries to create a diversity of employment that might take these communities
to a new level of sustainability.

PART ONE

Claudia Wong Fisher had moved to New York from the Midwest with her husband. They lived in an apartment in the heart of the city. His job was exciting, demanding, and paid well, but she was lost far from home with nothing to do. She had just graduated from college when she got married and had yet to forge a career path for herself. Her degree had been in business; her practical experience had been in her parents' gardening supply and nursery business, where she had worked while she was growing up.

Coming from a second generation Asian-American family, Claudia had been one of those self-starters who had organized gardening clubs in high school. She had started her own small catering business in college where she provided campus parties with good healthy food that looked like junk food; it was a huge success.

Claudia understood how to run a business and she knew how to grow plants. Growing plants was not only what she knew best, it was her passion. Somehow that didn't seem very practical here in the heart of the city. Land was at a premium and even a greenhouse would be frightfully expensive. One day at a party with her husband, Jack, she overheard someone talking about urban farming. They said that in the most devastated parts of the city empty buildings could be used for farming, and many empty lots existed where buildings had been destroyed. Claudia had found a way to get started.

Claudia's plan was to create an urban farm. She wanted to find an old building where she could convert the top story to a greenhouse and use the other floors for hydroponics gardening and maybe raising poultry. The bottom floor could be a produce and flower market and maybe a restaurant. It was an expensive and complicated undertaking. With the help of one of her professors from business school, who had many years of corporate experience before he began teaching, she felt she would be able to handle it. With his help, she was able to write up a comprehensive and convincing business plan.

After some looking around, Claudia found a land corporation owned by a church group that was interested in experimental projects that could help provide employment and stability in areas of high unemployment.

The company liked the plan because it offered a number of jobs to jobless residents near the site. Claudia made a list of the skills she needed and how many of those jobs would have to go to experienced people. She found a city job development agency that was willing to offer training while the building was being remodeled.

The building was on the edge of the South Bronx; it had been given to the church recently, and they had a study committee looking at possible uses when Claudia called. After reviewing her plan they decided to enter a limited partnership with her where she would have use of the building in return for a percentage of the profits. The corporation also helped Claudia obtain a bank loan for the funds needed to fix the building and start the business.

Claudia was persuaded to make the building as energy self-sufficient as possible; she redesigned the top floor so that there would be room for greenhouse skylights and solar energy panels. She was able to get Texas Instruments to donate some of their new low-cost solar cells, since she was providing job training and employment to currently unemployed people.

By using energy-saving devices at every step of the way and by personally supervising all of the work, Claudia was able to create an enterprise that was cost effective and created delicious vegetables. The top floor was used to grow salad greens in a greenhouse setting using soil and supplementing sunlight with grow lights. The third floor grew tomatoes and other vegetables in hydroponics solutions. The second floor housed chickens that were raised to be sold as fryers and were a source of eggs. The basement was used to compost all the wastes from the operation and to grow mushrooms.

One of Claudia's most significant management problems was keeping out the friends and relatives who wanted to see how the farming operation worked. The top floor greenhouse was everybody's favorite place, with its earthy smell and warm moist air. On the third floor the rows and rows of green plants sitting in their sand, gravel, or vermiculite bases with the nutrient solution constantly flowing around their roots fascinated the workers, who took great pride in making sure that the feeding solution was always just right. The chickens were a favorite of the children who came in and helped their parents collect eggs. The life and vitality of working with so many growing things infused an excitement and energy into the workers who were seldom late or sick.

The ground floor of the building housed a restaurant and produce shop. Claudia called the restaurant The Chicken Salad; it specialized in dishes from ingredients grown in the building. There was extra room on the ground floor, and a bakery producing whole-grain products seemed a natural addition to the project. Claudia leased it to a neighborhood group that paid its rent by providing bread to the restaurant and made a profit by what it sold to the public.

The urban farm's reputation spread quickly. As a result Claudia had people interested in opening similar operations. The problem was finding suitable sites. The building needed to be cheaper than New York real estate usually was in order to make a profit. But most of those cheaper sites were in places where most New Yorkers would prefer not to go.

Claudia organized a group of people interested in investing in additional urban farms. She knew that she did not want to be responsible for running more than one operation, but she liked the idea that her concept could spread, and she had learned so much in getting the first one started that she wanted to share that information. The church development group was able to bring in church groups from some of the inner city areas. The groups formed a new development corporation specifically dedicated to providing jobs in the inner city. The group decided that the South Bronx area could profitably handle an additional urban farm. A nonprofit corporation with its population basically from the church groups in the area was formed to build and run the urban farm. Claudia volunteered to work with the group as a consultant.

The newly formed nonprofit group began to work with several churches near the site for the new urban farm. They decided that they would trust the churches to pick some members who needed work. They also asked the church group to help find people who had become homeless and who would be able to work.

The response was overwhelming. So many people in the churches surveyed were looking for work that the group could not take everyone who wanted to be involved. Several people who were not chosen to be involved formed a committee to explore other oppor-

tunities. This committee started by asking what other
operations would work in synergy with the urban
farm. They came up with the following ideas: greenhouse mini-farms on some of the vacant lots; produce
stands to sell the food locally; mini home-based restaurants or bakeries to serve the local residents; a
weekly street market so residents could sell homemade items like clothing or crafts. They liked the
ideas they had come up with and divided into groups
to work on them.

Since Claudia's building was located in a potentially dangerous area where the crime rate was high,
she was interested in what the residents of the area
were doing and wanted to support their efforts. She
knew her business might not thrive for long if the tiny
pocket of prosperity that she was producing did not
have a chance to spread. The residents knew the
whole thing would be trashed if they didn't include
children and teenagers in their plans. Most of the kids
under sixteen had seen the effects of drugs and violence all around them and were open to other alternatives. But they weren't especially inspired by their
parents. As the MTV generation, they looked to music
and sports figures as their heroes.

One member of the church development agency
was asked if she would bring in a popular sports figure that she knew. When the projects that were under
way were described to the sports figure, he became
very excited and wanted to help. His participation excited the children, and the circle of those working for
change gradually began to enlarge.

By the time a six-month planning process had
passed, ten projects were funded and in place. The
new urban farm was under construction; additional
greenhouse projects were under way; several restaurants and bakeries had started in people's homes

(they had worked hard to bring their in-home operations up to the standards required to get a public
health permit); some home craft industries were
started; and the weekly crafts and food market was
about to begin.

The next critical step to making the planned projects successful was making the streets safe enough
that people could come to the markets, the restaurants, and the other businesses people were starting
to open. The core group of church members and citizens decided to do two things. They would get the
community involved in policing their own area and
they would ask for more police to help at critical
times.

The media loved the plan and sent in a few veteran
reporters who had already been working on the issue
in the South Bronx. They did in-depth stories on all
the new enterprises under way and then interviewed
the celebrities, the church groups, and many other individuals in the South Bronx. By the time the neighborhood market opened, most of greater New York
felt they knew the people of the South Bronx personally. Round one of urban farming and associated enterprises was a success.

Within five years a thriving infrastructure of urban
farming was located in the Bronx. The success of this
project had inspired both national church groups and
neighborhood groups to begin small farming operations in other parts of the greater New York area. The
intense interest in this activity generated some major
breakthroughs in low-light hydroponics farming that
allowed increased production. Entrepreneurs and inventors began to see small-scale food production as a
viable enterprise for their attention and investment of
time and money.

The people of the surrounding areas in New York

saw the people of the South Bronx get their act together and rebuild their lives. And they began to
think maybe it could happen in their own neighborhood. Claudia Fisher became very active as a consultant to this renewal process. She worked in trouble
spots around the country, and eventually around the
world, giving technical and business advice on how
to make urban farming a success.

PART TWO

*In this part of the story the change is from the bottom
up—gang members themselves help turn a bad neighborhood around while creating a sustainable new livelihood
for the residents.*

A spring morning in Los Angeles after a good rain
was a glorious sight. On this spring day, Emma Jones,
eighty-one years old, wizened matriarch of a Central
Los Angeles African-American family, was in her garden. She was talking to her plants and getting ready
for spring planting. Emma had always had a garden.
She grew greens and black-eyed peas to feed the bodies and souls of her children and grandchildren. And
she grew lettuce, tomatoes, potatoes, and carrots.

Emma was lucky. She had relatives living on each
side of her and years ago had talked them into letting
her use parts of their backyards as she continuously
expanded her garden. She used to have some chickens
until someone from the city had come out and said
that she couldn't have them here. Most of the food
that she grew went to relatives, and in exchange as
she grew older they helped her more and more with

the work. Although the truth was that at eighty-one, Emma could work harder and faster than any of them. She said the work kept her young.

Much of what went on in Emma's community confused and disturbed her. The drugs and the killings seemed to get closer all the time. Emma's grandnephew Cyrus was the leader of one of the most violent gangs in Los Angeles. Cyrus lived next door to her and used to help in the garden when he was little. Then his father died and his mother went on welfare. The morale of the family plummeted. Emma tried to help but the situation had gone too far. Cyrus seemed to need a lot of things that Emma just didn't have to give.

One spring morning Cyrus came home late, after the sun was already up, and saw Emma working in the garden. The simplicity and beauty of her actions were such a contrast to the sordidness of what he had just come from. He wanted to go out and dig in the soil the way he had when he was a little boy. He laughed at the thought. But a couple of hours later he found himself with dirt under his fingernails telling Emma about his current life.

Emma listened and heard the part of Cyrus that was begging for help. She said a silent prayer and asked Cyrus how she could help him. Emma said all she knew how to do was garden. And, if she could sell some of her produce she would be happy to give him the money, although she knew that Cyrus needed a lot more than money. But, Emma continued, the streets were so violent that she was afraid to go anywhere to try and sell her produce.

Suddenly, Cyrus realized that although he could go anywhere he wanted—he had guns and gang members to help protect him—people like his Aunt Emma couldn't. He sat on this thought for a few days. His

whole reputation with his gang was based on fear and
bravado; how could he talk to them about helping
little old ladies to cross the street?

Cyrus hadn't become the leader of his gang by being timid, and eventually he shared his ideas with the
other members. Some left. Some stayed and helped.
Within a few days Cyrus was at work building his
organization. Then he even began contacting other
gangs in the area, seeing who would become allies in
a program to make the streets safe.

A few months later a quiet movement was happening in the streets of Central Los Angeles. Cyrus had
helped Emma find a place to sell her produce and was
making contacts with other people with large gardens
to buy their surplus and resell it. There was hope in
the air. The number of shootings in the immediate
neighborhood had gone down, and people who saw
the success of the produce sales started to ask themselves what they could do to create a livelihood for
themselves.

The word of what was beginning to happen in Cyrus and Emma's neighborhood finally reached a city-
wide task force whose job was to bring business,
industry, government, and citizen groups together.
For years their suggestions and resolutions had not
led to action and ended up being filed away, perhaps
for some future archeologists to uncover and read.
But one smoggy October day one of the members of
the task force reported on the revitalization that was
happening in parts of Central Los Angeles. The other
members were skeptical but decided to make a field
trip to check it out.

One trip was enough to encourage several council
members to get together and try to figure out how
they could make these beginning efforts really take
off. They met with Emma, Cyrus, and some others in

the community who had started businesses, and they came up with some plans to help by providing money and technical assistance. They also worked with the city government to ease zoning restrictions that might have stopped some of the small businesses. Seed money from both government programs and private foundations began to filter into the neighborhoods that were rebuilding themselves. Local community groups formed to figure out what enterprises might be brought into the neighborhoods.

Since the original impetus to change had come from Emma's garden the officials looked at how that example could be used to provide role models for other enterprises. They brought out Claudia Wong Fisher from New York, who had been so successful in creating urban farms there. Claudia visited Emma and Cyrus and took a tour of the neighborhood gardens. She then recommended that they bring in a biodynamic consultant who specialized in intensively farmed outdoor gardens. She had a friend who had studied with the legendary John Jeavons who might be interested.

Emma wasn't familiar with biodynamics but she liked Claudia immediately as she sensed the bond they had in loving to grow plants. Claudia explained to Emma that biodynamic gardening involved using a small space with the soil dug and conditioned as far as three feet deep, much deeper than ordinary gardens. This process allowed the roots of the plants to go down rather than spread out and more plants could be planted closer together. And the constant mulching with organic material would keep the soil enriched so that ongoing crops could be harvested.

As a result of this new direction, gardeners in the community realized they could grow far more produce than they had been. Specialty gardens began to multiply, producing gourmet vegetables that were in demand by restaurants.

Cyrus worked with the task force to develop information on markets and specialty produce that would sell well. He found he had a talent for sales, and became quite successful at getting contracts with restaurants for selling the produce that was being produced in the neighborhood. But that only took part of his energy, and while his neighborhood was experiencing renewal he knew the job had just begun.

Knowing that gardening would not be enough to provide employment for the huge percentage of people in his neighborhood out of work, Cyrus began going to all the meetings of the planning committee and trying to recruit other businesses for his area. One day there was a representative from Pacific Bell who thought that some form of data processing and other electronically generated work that could be done anywhere would be a good place to start. He connected the neighborhood group with several large companies specializing in food and gardening supplies who had been thinking of moving their data operations out of the city and helped convince them to start a pilot project here.

The companies were reluctant to put expensive equipment into a poor and traditionally violent neighborhood, but one team member was so enthusiastic about the idea that they grudgingly gave it a try. At the very least the companies would get plus points with the media when their public relations people informed the press that they were setting up work stations and training inner-city people to use them. The results were not extraordinary, but the job got done and the facilities were not destroyed. More people from the neighborhoods began to approach the companies, and several new data processing centers were located in the neighborhoods.

Many of the new workers were women who had been on welfare and had small children. Transportation and child care were big problems for them, so small data processing centers were scattered throughout the neighborhoods where the women could walk to work. Soon child care centers were established next door to the data processing centers.

Cyrus meanwhile had become a full member of the council and was anticipating the next move he should make. He knew that the success of his neighborhood needed to be shared or it would attract jealous attention from gangs of other ethnic groups. It was a couple of months later when a new Latino American member of the council angrily approached the city planning group of the council asking why all the resources were being poured into African-American neighborhoods. Cyrus was expecting something like this and had a plan.

First Cyrus and the Latino council member met with some of the former African-American gang leaders and discussed how to produce the security necessary to begin the revitalization of the neighborhood. Many of the more conservative city development leaders were sure the process would break down here, since they didn't believe that the gangs could be brought under control, especially in the Latino community since gangs in the Los Angeles area had started in the Latino community.

But a truce was arranged with the most powerful Latino gang by appealing to their ethnic pride. They were taken on a tour of the neighborhoods that were making progress in revitalization and then taken back to look at their own neighborhoods. They were shamed by their elders for letting the other ethnic groups have the advantage in creating a livable environment for their relatives.

The truce gave the solid citizens in the Latino community time to organize and think about the kind of work they wanted brought into their community. Like the African-Americans they had always had gardens. Plans were made to expand and begin neighborhood markets. Several small manufacturing companies who had been considering sending their business to Mexico or the Orient were brought in and pilot programs were begun with small facilities.

The manufacturers studied what had worked with the data processing companies and created small-scale workshops with child care in the neighborhoods where people could walk to work. There was friction, and several times rival gangs damaged the new workplaces, but the people in the neighborhoods were determined to keep the work they had brought in. They formed neighborhood patrols to protect the new workplaces and the gang attacks stopped.

Encouraged by their success in protecting the workplaces, the neighborhood patrols expanded to include wider and wider areas. It took time but eventually the women and children of Central Los Angeles could walk alone on the streets at night.

With some financial security the people in the African-American and Latino communities began to take control of their lives. And before the rest of the world knew what was happening a new garden spot had developed in the middle of what had previously been one of the most crime-ridden, degenerate areas in the country.

The media had a brief heyday when they discovered the changes that had occurred. Suddenly Central Los Angeles was a tourist attraction. As with most fads it was soon out of the news, but the national attention left the residents of the neighborhoods with new ambitions and desires to continue the transformation of their neighborhoods.

Neighborhood schools were revitalized from within as parents who now had incomes and pride began to volunteer time and money to make the schools better. They volunteered in the classrooms, supervised homework, planted flowers, painted walls, and replaced broken glass.

Those people still on drugs or in gangs began to feel more and more isolated. Some moved away to areas that were still crime-ridden, and some gave in and became productive citizens.

The miracle that started with Emma's garden began to spread to the rest of the city. And people who had moved out of the city because of fear of crime began to consider moving back in and getting away from the long commute.

The success in South Central LA made all the previous efforts at creating change in the city reexamine their tactics. A task force on redevelopment was formed, composed of many of the government and business leaders who were on many of the previous councils and task forces, but this time they hired Cyrus as a consultant.

Emma Jones died in 2003 at the age of ninety. Cyrus became a successful businessman and in his later years a dedicated politician.

10.
A New School

*It is impossible for us to visualize a move to a sustainable
society without thinking about education for our children.
Children need to learn why cooperation is necessary, how
joining their efforts together can make a stronger group,
and how all the parts of a sustainable society works. The
main character in this story, Hannah, has been caught
between two cultures and is searching for the best in each
to pass on to the children in her care. The school she creates is full of opportunities for children to learn about
their place in nature.*

The school of the future first appeared in the
countryside of Pennsylvania, a place where the
past and the present struggle to come to terms
with each other. The simple lifestyles and ages-old
beliefs and practices of the resident Amish farmers
ground uneasily against the modern lifestyles of
many neighbors.

By the mid 1990s it was becoming more and more
clear that the traditional ways of the Amish had many
things to teach those interested in sustainability. The
Amish farmers had never stopped composting; their
produce was organically grown. And they certainly
believed in recycling and reuse, although they would
simply call it "not wasting." They did not use sol-

vents and toxic chemicals in their crafts or food processing. In the small towns where the Amish shopped
and sold their wares a new level of respect from outsiders was developing.

In these towns there were always a few people who
had left their Amish upbringing and settled in with
the outsiders. One midsized Pennsylvania town had
enough of these people to form a small community,
and one of the women in that group started a school.

Her name was Hannah Goodman and she had been
raised until the age of sixteen on an Amish farm. Hannah had a happy childhood in the Amish community,
but she enjoyed the barn raisings and the corn thrash-
ings more than the endless chores she was forced to
do inside. She loved the early morning when she
could walk from the house to the milking shed just
as the sun was rising. It was a shock to her when her
family moved to the city and the noise of cars com-
peted with the birdsong as she awoke.

Her parents had been forced to leave when they
had a philosophical disagreement with the elders of
their community that they felt they could not back
down from. They were people of high moral character, but had come to believe that some change was
inevitable. They were in favor of a careful inclusion
in their lives of goods and services used by the rest
of the world.

They, of course, did not approve of the moral attitudes they found outside the Amish community. Pre-
marital sex, drugs, rock music, and violence all
seemed foreign and evil to them. But as they looked
around them in the community they found other responsible and hard-working citizens and they felt at
ease in their company.

When they moved they opened a shop and sold
crafts that they made or obtained from sympathetic

Amish in some communities farther away. They expanded their line to include Shaker products and others of similar craftsmanship. The shop was successful, and the Goodmans decided to let Hannah attend the local junior college.

Hannah's views were already very liberal by comparison to those of the traditional Amish. Her exposure to the college environment further broadened her perspective. In her parents' shop she had a lot more contact with outsiders than other Amish. She was exposed to material goods and the technology that seemed to make life easier without doing any harm, and she met many good people who were not Amish.

At first, Hannah wanted to learn more about business to help her parents. But she found much of the course information irrelevant, lacking in the good, solid, common sense that seemed to come to her naturally. She also took some education classes, since her other desire was to learn to be a teacher. But like the business classes, she found the classes in education lacking in what she felt children needed to know.

However, these years in school exposed Hannah to more than classes. She met other students and found friendships in a group that organized wilderness trips. There she gained self-confidence and an even deeper love of nature than she had as a child.

Even though she did not always like the classes, Hannah was the kind of person to finish what she had started, so she continued to pursue her degree in education. She found her views so different from what she was learning in school that she came to believe the only way she would be able to teach was in a school of her own. And she thought it would be easier to open such a school if she had credentials from the state.

Hannah was ready to open her school the fall after

she completed her education credential. Many of the former Amish in the community wanted a school where the children would not lose all of their old ways. Since no one had any money to finance an elaborate setup, the school was started in the basement and backyard of one of the families, and for the first two years it was just Hannah, with a few parents helping. The parents were so happy with the results that within two years the reputation of the school had grown and not only Amish people, but other members of the community who were looking for alternatives, had put many children in the school. Hannah could then afford to hire a second teacher and move to a house where zoning for a school was allowed.

Hannah had already decided that much of what existing schools did either wasn't necessary or didn't work. Children spent a lot of time pretending to learn facts that they weren't interested in or seemed irrelevant to them. They had to relearn the same material every year when school started in the fall, and the ultimate application of almost everything they learned was questionable.

Hannah and the parents she was working with decided first on some general goals; they wanted the children to find a balance between the traditional and ecologically sound ways of the Amish communities that most of them had belonged to, and the technologically advanced world that seemed to be everywhere else. They wanted the children to be able to make good decisions about which advances were useful and which were destructive to the environment. They wanted their children's education to have many of the practical elements of their own without the rigid dogma. All school subjects should be taught in context of these objectives.

Hannah gave the school its initial form by having

a garden as the focus of activities. Each child had a job in the garden and was patiently shown how to do it. The mornings were spent working in the garden and on crafts like carpentry. The children were taught to cook and fix their own lunches. "Be aware of where the things you need to live come from" was a major theme.

At first most of the food for lunch was brought in by the parents, but as the garden became more established food for lunch began to come from the garden. What was grown changed over the years, but it was always organic. The property that the school was on had fruit trees and after a few years, beehives were added. The older children were taught to handle the bees, a job that almost all of the children were eager to learn. Finally chickens were added for their eggs. The children could then prepare a fairly well-rounded lunch from their own resources.

Hannah wanted the children to know that all of their food came from somewhere before the supermarket. To obtain the supplies that the school did not grow they went directly to the producers. Hannah and the children went to a farmer for wheat and oats, which they bought whole and ground when they needed it. Milk came from a nearby dairy where they helped to milk the cows. If they wanted butter they had to churn it themselves. To the Amish this seemed natural, and they did not want their children to lose these skills.

Running the farm and cooking lunch was a lot of work. And the school was only successful because of intensive parental involvement. After the school had been running for a few years many of the older children could take over some of the supervision of the younger children, but parental involvement was still needed. Most of the parents decided that if they needed to give a lot of time to the school they would enlarge the garden so that it served the home needs of the school community as well as the day-to-day school needs.

As they got older the children learned wilderness survival skills, and from the sixth through the eighth grade they went on annual camping trips. They were taught to identify plants they could eat for survival, how to build temporary shelters, and how to find their way out of a wilderness area if they were lost. But most of all on these expeditions they learned how humans are dependent on each other and on the fertility of the earth to survive.

Hannah had observed in her education classes that children were seldom given much practical knowledge about how to take care of their bodies, including knowledge of how the body works and how to prevent and treat simple illness. So, in her school even the youngest children were taught the importance of nutrition and how to clean a cut or scrape and put on a bandage. By the eighth grade some students could qualify as barefoot doctors (the medically trained civilians in China in the 1960s).

The importance of diet, exercise, and mental health were not abstractions but were part of working in the garden and learning to prepare healthy and delicious foods. A few common medicinal herbs were grown and some common wild ones were identified, so that an eight-year-old with a tummyache might go out and pick some chamomile flowers and make herself a cup of tea.

All children learned to take their pulse, keep charts on their weight and height, and keep a diary of how their bodies felt under all kinds of circumstances. The children learned which nutrients were present in what foods so that they could always put together a healthy combination of foods to eat. And they were taught to report any unusual feelings or pains to the teachers or their parents.

The Amish traditionally rejected TV in their homes and never went to movies, but Hannah got the parents to see the value of a VCR as instructional material. She also was able to get them to accept the use of computers in the classroom when she showed them how gentle the learning process was for the children. She felt that computers allowed the correction of mistakes by the students without the overtones, even of a patient person, telling the student that they are wrong.

Hannah learned to use the best of the new technology to ease her students into basic skills like reading, writing, and arithmetic. Math, science, geography, history, music, art, and English literature were all taught within the context of what the child was interested in at the time. Videos and computer programs were as frequently the source of information as books.

The core of Hannah's program was to help each student find their individual strengths, and to learn how to work with others. Teamwork and cooperation were taught in the gardening and the cooking where the children learned to work as a community. And the children were encouraged to express their individuality by creating their own projects on subjects that they were interested in.

Hannah encouraged whatever interests the child had and taught them how to research their interests with the computers and libraries available at the school. Reading and writing were taught to the small children as ways to tell stories or to gather information on the subjects they were interested in. Small children who loved stories, fairy tales, and animals would write books or make up plays.

As they grew the children were encouraged to continue their individual projects. And field trips were frequent for the middle children to visit places of work, and see what people actually did for a living. The older children were encouraged to pick areas of work that interested them and spend time with people doing that job.

It took a lot of time and effort to line up suitable, willing people who would allow the children to visit and/or work with them. But, again, this worked because of parental involvement. Many parents volunteered and got their friends and colleagues to volunteer.

The children were encouraged to set up small-scale businesses, and everything from selling lemonade to investing in the stock market was tried. The teachers in consultation with the parents tried to identify the skills and interests of each child so that they could help steer them to the necessary resources for these projects.

Hannah, of course, always wanted to bring her students back to an awareness of their place in nature. All lessons on business included a look at where in nature the raw materials came from and where they would go when they were no longer useable.

Hannah also believed that laughs and hugs were essential to every child's well being. And the children were taught that a day without fun was a day wasted. This concept did not come from her Amish background, but was one of the concepts, in addition to the use of VCRs and computers for teaching, she had picked up in school and that she felt was worth adding to her curriculum. Left to their own devices most children will naturally laugh and play, but Hannah wanted them to remember it when they got older, like brushing their teeth or taking a shower.

Hannah inspired a revolution in education but few people outside the field of education ever heard of her. Her emphasis on common sense skills and the development of individual talents changed the field of education and helped a generation of people become flexible and creative. She showed that teaching cooperation in group efforts led her students to a greater ability to understand the need for sustainability in society. She eventually married and had her own children, and she continued to run the school and make improvements for the rest of her life.

11.

Security and
Sustainability

*Anyone who has watched the scenes of military conflict
on TV knows that war is hazardous to the environment.
Now the military, who in the past were only taught to
wage war, are being given new responsibilities as peace-
makers and peacekeepers. Now in order to discharge that
role we must cooperate with other governments and their
military organizations. The underlying role of the military has always been to create security, which used to
mean keeping the country safe from an attack by an outside nation or force. Some people see new threats to our
security in the environmental destruction occurring
around the world, and that a proper goal for the military
is helping to alleviate environmental problems. In this
story we see the military wrestle with this new role and
to try and understand its role in relationship to long-
term sustainability.*

In the late 1990s the American military began to
redefine itself. The Pentagon and the joint chiefs of
staff announced that the new goal of the military
was creating security. They went on to explain that
identifying the causes of conflict and trying to diffuse
the conflict before it happened would make the world

more secure, and that many issues that threaten the country were environmentally related.

The current head of the joint chiefs of staff, General William Garcia, was a veteran of fighting in Somalia and Serbia. He had seen the end of the Vietnamese war when he was first commissioned. He had been involved in every conflict from Vietnam to the Gulf War with Iraq in 1991. He had seen overt and covert action, and had gained his reputation by being a thoughtful, thorough commander who had never asked his troops to do anything he hadn't tried himself.

William Garcia grew up in a family from Mexico in San Antonio. He had a sister who was a nun and an uncle who was a priest. His dream from childhood was to go to West Point. He was an idealistic young man who believed in keeping the world safe for democracy. His first assignment was in Vietnam, where his platoon saw a lot of front-line action. He had killed men, and seen women and children die in the line of fire. Vietnam was not his idea of what war was supposed to be about. William had seen wholesale devastation of the land in Vietnam, and he felt great regret and discomfort at what he had witnessed and at his role in the conflict.

Later in his career he was involved in Somalia and other relief-related actions. The faces of desperation and starvation became more familiar from these operations than the faces of a well-armed combative enemy.

In the years following Somalia and Bosnia, William was responsible for much of the soul-searching that went on in the Pentagon. William truly believed that the job of the military was to protect. It was this conviction that kept him going throughout his career. When he was younger he thought his mission was just to protect American lives, but after his years in the field he had come to believe that he could not protect American lives without trying to protect the ecosystem which gave all humans life.

William and the other men in charge had not been trained to take care of people. They knew about nuclear bombs and limited warfare over resources, but to be placed in the role of fighting those who would steal food from their own people, or placed in the midst of centuries-old blood feuds, was unnerving. The combat troops assigned to these missions burnt out much faster than in regular combat situations. The constant sight of starving and wounded children was more than many could take.

The innocent had always been hurt in wars, but for soldiers trained for an increasingly automated long-range battle, to have to engage in hand-to-hand combat where it was impossible to get civilians out of the way was painful and disturbing. Population increase in the third world countries made scenarios like Somalia more and more likely. Continued attacks in this country made a new national strategy necessary. And ethnic conflicts after the breakdown of the Soviet Union could go on for decades. The military began to ask what it should be doing in these cases.

The war colleges where young military officers were trained to plan strategies had new assignments. They needed to understand how to deal with famine and fighting within a country, how to deal with terrorist attacks, as well as how to attack and destroy the enemy. They began to try to understand ways to avoid conflict, and to look for examples of how conflicts were resolved without fighting.

William Garcia headed a task force that came to the conclusion that to avoid conflict people needed 1) enough food, water, shelter, and clothing; 2) feelings of cooperation and working for mutual benefit with neighbors; and 3) ways of distributing excess to those less fortunate to prevent jealousy.

They also concluded that for internal security in the US, so that terrorist attacks would not disrupt vital services, these services should be decentralized. This decentralization included power production, food growing, food storage, and moving away from a dependence on foreign oil.

The military began to advocate the development and use of energy from renewable energy sources such as the sun and wind, and to advocate decentralized development of these energy-generating sources. Sustainable, renewable technologies were the best thing for security as well as a stable ecosystem.

They further concluded that the more self-sufficient each household could become, the more secure the whole country would be. If each household produced all, or part, of its energy, and if each household maintained enough food and emergency supplies for long-term survival, then natural disasters would have less of an impact.

William Garcia saw that most of these areas were not within the traditional scope of the military, whose job was always seen as protecting a community from outside invasion. The military could make these recommendations, and use them for their own installations, but it was up to governments to legislate or encourage their use.

The scale of immigration from third world countries was becoming a kind of outside invasion to this country, and troops were called out at times to help on border patrols. It seemed that a better way to stop the immigration was to create security for these people where they had come from. This had to involve not only guns to protect these citizens from dictators

and warlords, but to somehow help people start to
create a life in places frequently overpopulated and
devastated of plant and animal life.

People trained in humanitarian efforts had not been
able to adequately distribute supplies needed to prevent the breakdown in Somalia. One general jokingly
said that what was needed was a well-armed Peace
Corps: people trained in ecological restoration, population control, and conflict resolution with big guns
to back them up.

The US military saw that some of these goals might
be best accomplished by working in cooperation with
the military from other countries or with a UN peacekeeping force.

One of the things William Garcia would be remem-
bered for was that he had the courage to try what he
and other military theorists were proposing. In the
late 1990s, a first operation that emphasized establishing security was tried in a small African country that
was suffering from the same kind of breakdown that
had occurred in Somalia a decade earlier. A group of
troops accompanied by UN forces went into the country with guns, tanks, food, contraceptives, and personnel trained in conflict resolution. Their strategy
was to move slowly through the countryside, leaving
secure and well-armed feeding and rehabilitation stations behind them.

The warlords who had terrorized the country and
contributed to its collapse were invited to a meeting
where possible futures were laid out for them. After
a few days they began to get the picture that they
were not going to be left alone to pursue their former
reigns of terror.

Ultimately the operation was a mixed success. It
was clear that developing a new ethic on the part of
the warlords would take time and would require the

continuing presence of outside military. Troops were left and the work of making the country self-sustaining and free from terror continued.

In the meantime it became clear that the resources of one country, namely the US, were not sufficient to meet all the crises present in the world. The US and the UN then set up task forces to solicit more coop-eration from the wealthiest European and Asian coun-tries to set up larger UN peacekeeping forces.

While William Garcia was deploying forces to third world countries and working on trying to establish an effective international peacekeeping force, a brilliant young Japanese economist wrote a book entitled *The Economics of Peace*, which was translated into at least fifty languages within its first year of print. In it he showed, using complicated mathematical formulas and common sense applications, that the world econ-omy in fifty years would be ten times wealthier if worldwide peace and sustainable energy use could be attained. The methods and conditions of worldwide peace that he outlined were much the same as the US military task force had come up with a few years ear-lier.

By the year 2000 a real momentum had started to build with business, government, and the military starting to work together for commonly stated goals: economic sufficiency for all people, environmentally sustainable practices, and population reduction, which would help the previous two to happen. By the late 1990s even religious groups that opposed birth control had begun to admit that maybe humans had fulfilled the order of God to multiply and fill the earth, and were assuming more open attitudes toward birth control.

By 2005 the look of the military had begun to shift. All members of the armed forces received training in first aid, agriculture, and conflict resolution. When they were sent into a trouble spot it was in multi-ethnic teams including individuals from the area's ethnic group.

By 2005 some of the global-wide ethnic strife was being brought under control. The global commitment had switched to a movement to clean up the ravages of the industrial age and create a sustainable future for all people on earth.

When William Garcia eventually retired from the military, he quietly joined a religious community, and spent the rest of his life trying to find inner harmony. He told his family that what he had seen in his lifetime had left him spiritually drained, and he wished to spend his final years in peace and tranquillity.

part IV

ENVISIONING
A SUSTAINABLE
SOCIETY

Human societies are the result of dreams and visions. Different forms of government, economic structures, and religious beliefs determine how we distribute goods and services and how we treat each other. And, as varied as these forms are, they all came from the human mind.

Our environmental problems are the results of dreams and visions that came in conflict with the workings of nature. Our dreams did not take into account nature's patterns and limitations, and we did not design our products and processes with these in mind. A sustainable future involves having a new vision, one in which our needs as a society are seen in the context of what nature has to give.

What would that dream look like? One of the most important elements is a belief that the needs of nature are as important as the needs of humans. That means that we have to give nature the space and opportunity to function. Ecosystems such as rain forests cannot be completely destroyed and reseeded at our whims. We might move to selective harvesting to preserve the forest and give income to its inhabitants.

If, in a sustainable society, the wood products we need came from selected trees in a forest, then when it becomes clear that wood obtained selectively will be more expensive, then we will have to examine all the ways we use wood. Maybe other resources can be found to be cheaper and more sustainable for many of the things we now make out of wood. The same line of thinking is true of everything we make and do. How does this activity or this product fit with the working of nature? Once we start thinking that way the creative human brain will come up with solutions to our environmental problems that will astound us, solutions we really cannot predict.

If you look at any writing that seeks to project the future, you will find that no matter how imaginative the idea, the technology is a linear projection of what we know how to do. H. G. Wells's future projections involved mechanical devices based on the technology of his day. And even though the science fiction movies of our own era involve movement across distances that we have no idea how to achieve, the interiors of their ships are usually familiar, involving technologies that are somewhat similar to ours.

But go back two hundred years, to the time America was settled by Europeans. Think about the shock a person from that era would experience on a visit to one of our major cities. The sights, the sounds, the feel of the place would be beyond anything they could have possibly imagined. Airplanes, automobiles, telephones, television, the way people dress would all be incomprehensible.

The world that will exist in two hundred years is incomprehensible to us. But that world will be based on the dreams and visions that we begin to create now. America today is based on the dreams and visions of the frontiersman of two hundred years ago. At that time the frontiersman dreamed of a farm to feed his family and a large family to take care of him in his old age. He saw the land he was settling as a wilderness that needed to be tamed. Multiply that dream thousands of times as people fanned out across this country. Within a short time, the forests and prairies were completely transformed by people sharing that dream.

Most of that wilderness is gone, replaced by individual farms, factories, cities, and suburbs. The same dream is fueling the continued elimination of wilderness in other countries around the world. A place to farm or raise cattle to feed a growing family has always seemed more important than the preservation of savannas, tundras, or rain forests.

A sustainable society must first of all dream fewer people. Only in industrialized societies where material needs are met has that dream taken hold of the imagination of the people. Regardless of religious belief or ethnic background, most successful, fully integrated members of American, European, and Japanese society have only a few children. More women are educated and their energy often goes to having a career and making money, and they cannot care for many children. Health care and nutrition for infants is good and it is likely that the children they do have will live to adulthood. The dream has changed. As that dream spreads around the globe the world will produce fewer people.

We are also now beginning to dream of harmony with the natural world. More and more people are interested in ensuring that all species of plants and animals have a place in our environment. That is a very different dream from the frontiersman who sought to tame all areas of the wilderness, including getting rid of any wild animals he encountered.

We dream of a world without the pollution and destruction of natural resources created by the industrial age. We dream of clean energy and the reuse of materials so that what we make and use has less impact on the land around us. This is a new dream and maybe we can pass this dream on to the undeveloped parts of the world so they won't destroy their natural resources before they find out what they have lost. The very idea of a sustainable society is a new dream. We only learned recently that our supplies of raw materials could be used up. We only learned recently that we needed to work at taking care of what we have.

As we end the journey that we started by looking into the inner workings of the cell, let's take a final trip into the future with a story of what a possible dream of sustainability might look like.

12.

The Sustainable Future:
A Possible Dream

The example earlier in the book of the ultimate sustainable society is a climax community—a place where energy and materials are constantly renewed by natural forces, or recycled and reused. As we look into the future in this story we find people still striving for a climax community. They have reached a point where much has changed and the world has attained a new balance, a balance good enough we could call this a sustainable society.

SIXTY YEARS IN THE FUTURE

Maria Jensen was looking forward to her installation of a new playmeadow. Maria creates architectural environments for schools, homes, offices, and places of recreation such as restaurants. She works in a field that evolved during the last fifty years as a synthesis between art, interior design, and landscaping. Today she is putting in what could be called a living carpet at the local preschool. Made of tough, resilient grasses and clovers, it thrives on the rough and tumble of preschool children, and helps keep them calm.

Maria loves working with natural, frequently still-growing materials in an architectural context. These materials have begun to replace the synthetics so popular in the last half of the twentieth century. With organic methods, the growing of carpets, wall coverings, and sometimes supports for the buildings is one of the most environmentally friendly industries to have evolved.

Maria prefers to grow her carpets in place, but the preschool was in a hurry to move into their new building, so she is overseeing its installation today.

This morning her husband Bob told her he would definitely be going to San Francisco tomorrow for the Entrepreneurs Conference. Bob Jensen works on the research and development of new businesses and has some contacts he wants to follow up on at the conference.

After a very quick breakfast (some things never change) Carlos and Sara, the Jensen children, slip their computer disks into their backpacks and pedal away on their bicycles. Most schools are close enough that children walk or ride their bikes. Mr. Jensen walks to the neighborhood work center today. Mrs. Jensen spends some time in her home studio before she gets ready to go to lunch with some business associates.

Bob used to work at home but he likes the companionship of other workers, and when a space opened in the neighborhood work center he jumped at it. Now he has someone to take coffee breaks with and gets to bet on the "office" sports pools. He does traveling to keep up on the latest business developments, but most of his research is done by using virtual-reality telephoning. It is, of course, not the same as being there in person, but it is much cheaper and the way most long-distance communication is done. It has greatly cut down the need of executives to travel. And Bob enjoys being home with his children.

In addition to the living materials, Maria works with fabric, lights, sound, smell, and something she calls harmony. Her goal is to stimulate the imagination while allowing the mind to be clear and focused. Most of her work imitates settings in nature, and currently she is working on a room to feel like a redwood forest for a corporate headquarters in San Francisco. Using water is a current favorite of hers and she is planning a small creek to run through the forest.

Maria is currently interested in research that shows that air and water from natural settings have more "vitality" than air and water that have been through any kind of human process. She is trying to incorporate this knowledge in her designs by installing air and water conveyances made in spiral shapes to simulate its natural flows.

She does her best work when the children are at school and Bob is at the office. When the house is quiet she feels she can do in a morning what it used to take her all day to do when she worked in a group studio. Today she is having lunch with a client. Then this afternoon she is going to spend some time in Sara's class.

The schools request at least an hour a week from each parent of each student. And with most parents working close to where they live, it is a request that is usually met. The schools ask the parents to do a variety of things; yesterday Bob taught his son's class how to make compost in the garden.

Without a commute and with control over their own projects, a six-hour work day is common. Bob and Maria are usually both home after school, when the whole family spends some time working in the greenhouse or garden, then they all help fix dinner.

Bob has to go to a conference and will be staying

with his sister in San Francisco. He takes the high-speed magnetic levitation train which rides on an air cushion above the roadbed. Most of his business travel is on trains, which because of their new accessibility and speed have become the preferred means of traveling for distances up to one thousand miles. Because Bob and Maria's work and family are within a thousand-mile radius, they have only flown once.

As Bob travels he looks out the windows at the multitude of wooded areas that house a large number of nature preserves. He feels grateful at the work his parents did in working for restored habitat for many animals in the area.

Bob gets to his sister's apartment in the city late in the afternoon. She has added him to her door voice lock. He says his name and walks in. Bob's sister, Mary, is a teacher. She is single and has a teenage daughter, Tina. To celebrate Bob's visit the three of them decide to go to dinner in a new restaurant downtown. It is called The Greenhouse, and the kitchen and eating area are in the middle of a huge greenhouse where much of the food served at the restaurant is grown.

During dinner Bob and Mary talk about how the city has changed since they were children here. Population decline has opened up space and made places like The Greenhouse restaurant possible. Pollution control has made the fish in the bay safe to eat, and many overfished species are returning. The control of both pollution and crime has made a beautiful city like San Francisco glorious. As they walk back to Mary's they comment on how thick the trees have become in the city. With more open space and two generations before them actively planting trees, more and more of the city is becoming a park.

The next day Bob goes to the conference on emerging business opportunities. His specialty is spotting trends and helping innovators and inventors get venture capitol for their enterprises. He listens to one presentation on the newest trend in environmental cleanup, bioreclamation, a mining process using natural bacteria forms to transform waste, mainly from old landfills, back into its raw materials. He is interested and files it away for future reference.

Bob's new client is an African-American woman who is marketing a fabric made of an African plant being grown both in Africa and the American south. It produces a fiber which, when made into cloth, is soft, durable, breathable, and can be grown in brilliant colors. Undyed clothing has become very fashionable, but most of the colors are muted. Brilliant natural colors would be a big seller. Bob is able to spend a lot of time with her and they work out a contract and a marketing plan for her fabric.

Later that evening back at Mary's they all watch a TV special on hunting preserves. After years of conflict between animal rights activists and hunters there are signs of a truce. Hunting has become a much more rigorous activity. Gone are the drunken orgies that stereotyped the hunter. Only the most challenging weapons are now allowed for hunting, like bows and arrows and single-shot muzzle loaders. Hunters are not given permits until they prove their ability to make a clean kill using virtual-reality testing. There is still controversy, but the animal rights activists who pressured hunters in the past have admitted that the care and reverence of the modern hunter gives them much to think about.

Bob and Mary get into a heated discussion after the show is over. Mary has been a vegetarian from her early twenties and doesn't believe in killing anything. Bob still eats meat, although much less meat than people did in the twentieth century, and said that hunting was a more honest way to get meat than at the supermarket. Tina said that they were both being silly and told them about a presentation in one of her classes that day.

Tina had seen a presentation on how the breeding of livestock and people's eating habits had changed during the last fifty years. Factory farming had been eliminated and the price of meat had gone up, causing a greater decline in eating meat than had the earlier reports on the health problems caused by eating meat. New sources of meat and poultry have come from many small-scale operations that are able to produce meat more humanely, and with higher prices for their products many new jobs have been created. Most open-range cattle herds have been replaced by buffalo or deer herds appropriate to the habitat that produces meat with much lower fat and cholesterol.

After a successful business trip and an enjoyable visit with family, Bob returned home.

Community life is strong in the town where Bob and Maria live. There are many festivals and celebrations during the year that bring people together and help to cement the community bonds created by living and working together. Bob returned home in time for the fall community produce and seed exchange, swap meet, barbecue, and talent show. It is a time for people to show off their produce and just about everyone brings some of their home-grown organic produce to exchange. Many people are into home gardening and quite a few enjoy growing unusual heirloom plants or produce their own hybrids. Little towns like the Jensens' could show Luther Burbank some startling finds when it comes to plants. Carlos brought a spectacular array of multicolored Peruvian seed potatoes to trade.

People came with everything they didn't need or want to be sold, traded, or just given away at the community center and its surrounding sports fields. The Jensens got rid of some clothes and books and picked up some pots for indoor plants. Later at the talent show Maria sang and Sara played keyboards for her. Bob and Carlos cheered.

The talk all day had been about the president's controversial plan to reopen immigration. For fifty years the US and other developed nations had all but sealed their borders to immigrants. But after half a century of expanded use of birth control and third world economic development, the population of the world was actually declining. Instead of peaking at ten billion as many experts predicted, population had briefly hit six and one-half billion and quickly gone back down to five billion. Since then it had declined to four billion and now experts were predicting population would stabilize around three billion, although some people thought it might go as low as two billion.

People had come to realize that a lower population base gave everyone a higher standard of living. But people never seem to lose their love of arguing the fine points of a juicy political issue, so the debate was heated and continued over many cups of coffee and tea and glasses of wine as the evening wore on.

The next day the Jensens went on the biannual one-day retreat for their church. Religion had added a new element in the past fifty years. The new players were earth-centered religions, and some had started churches. The Jensens belonged to a church that was still nominally Christian, but had adopted rituals and beliefs from Hinduism, Buddhism, and the new earth religions. The church advocated meditation, personal spiritual commitment, reverence for nature, and a belief that Christ was one, but not the only, representative of God on earth.

The retreat was done twice a year to help people seek personal spiritual fulfillment, which more and more people were seeing as essential to healthy living. After the science-centered twentieth century, the twenty-first century came to see values, dreams, and spiritual connection as important and maybe more important than scientific fact. Most people now belonged to some formal religious or spiritual organization and took its messages seriously. Bob especially missed not having some expression of faith in his childhood and made sure that Sara and Carlos had a church to belong to.

The retreat was held in an ancient forest close to the town where people lived. The dimly lit open spaces beneath the venerable trees was a perfect place to try to bring body, mind, and spirit into harmony. This fall's retreat dealt with how our dreams today are the reality of our grandchildren and asked participants to seek higher guidance so that their dreams would truly be of benefit to their descendants. And, so we leave the Jensens, dreaming their grandchildren's future.

There is no doubt that what we do today will determine the quality of life our children and grandchildren will have. Learning the patterns of nature and seeing how to use nature's resources can establish a whole new technological system. That is an important step toward a sustainable future. But just as important is maintaining our beliefs and hopes for the future. Nature shows us the way and our dreams and visions give us the tools and the hope to get there.

SELECTED BIBLIOGRAPHY

"All Tired Out." *People*, August 1991, 57.

Amirrezvani, Anita. "Recycling Computers, New Life for Old PC's," *Computer Currents*, July 1993, 46–47.

Anderson, Edward J. *Transit Systems Theory*. Lexington, MA: Lexington Books, 1978.

Anzovin, Steven. "Eco Logic." *Compute*, July 1992, 82.

Appelhof, Mary. "Compost Indoors: Worms do the Work." *Organic Gardening*, January 1992, 58–63.

Baer, Steve. *Sunspots*. Seattle: Cloudburst Press, 1979.

"The Bell Solar Battery." *Bell Laboratories Record*, Courtesy AT&T Archives, July 1955, 241–246.

Berger, John J. *Restoring the Earth*. New York: Alfred A. Knopf, 1985.

Blair, William. *Katherine Oreway: The Lady Who Saved the Prairies*. Washington, DC: The Nature Conservancy, 1989.

Blumenechine, Robert J., and John Caballo. "Scavenging and Human Evolution." *Scientific American*, October 1992, 90–96.

Braasch, Gary, and David Kelly. *Secrets of the Old Growth Forest*. Salt Lake City: Peregrine Smith Books, 1988.

Bullock, Charles E. *Solar Electricity: Making the Sun Work for You*. New York: Monegon Ltd., 1981.

Bushsbaum, Mildred and Ralph. *Basic Ecology*. Pacific Grove, CA: The Boxwood Press, 1972.

Calem, Robert E. "Working at Home for Better or Worse." *The New York Times*, 18 April 1993.

Calthorpe, Peter, and Sim Van der Ryn. *Sustainable Communities*. San Francisco: Sierra Club Books, 1986.

Carrying Capacity Network: Clearinghouse Bulletin. 1325 G Street NW, Suite 1003, Washington, DC 20005–3104.

Chadwick, Douglas. "Denali, Alaska's Wild Heart." *National Geographic*, August 1992, 63–87.

Christensen, Kathleen. "Remote Control: How to Make Telecommuting Pay Off For Your Company." *PC–Computing* 3, 90–95.

Clark, Desmond, *et al*, "The Last Stone Ax Makers." *Scientific American*, July 1992, 88–93.

Coop America Quarterly. 1850 M Street NW, Suite 700, Washington, DC 20036.

Colinvaux, Paul. *Why Big Fierce Animals are Rare*. Princeton: Princeton University Press, 1979.

"Conservation Power." *Business Week*, 16 September 1991, 86–92.

Cramer, Jerome. "The Selling of the Green." *Time*, September 1991, 48.

Cunningham, Scott, and Alan L Porter. "Communications Networks." *The Futurist*, Jan–Feb 1992, 19–23.

Deja Shoe Press Packet. Environmental Communication Associates, Boulder, CO.

Duddington, C. L. *Beginner's Guide to Botany*. New York: Drake Publishers Ltd., 1970.

Erlich, Paul, and Robert Ornstein. *New World, New Mind*. New York: Doubleday, 1989.

Evans, Ianto. *The Permaculture Activist*, August 1992, 25–29.

Fischetti, Mark. "Here Comes the Electric Car." *Smithsonian*, April 1992, 34–44.

Fisk, Pliny. "The Ecological Business." *Center for Maximum Potential Building Systems*, Austin, TX, 1989.

Folsome, Clair Edwin, ed. "Life, Origin and Evolution." *Scientific American*. San Francisco, WH Freeman & Co., 1975.

Gore, Al. *Earth In The Balance*. New York: Houghton Mifflin, 1992.

Herndon, Jay. "Ten Years Without the Grid," *BackHome*, Summer 1991, 28–31.

Hillel, Daniel J. *Out of the Earth: Civilization and the Life of the Soil*. New York: The Free Press (Macmillan), 1991.

Holden, Ted, and Richard A. Melcher. "A Trading Floor On Every Screen." *Business Week*, 5 November 1990, 128–132.

Horton, Tom. "Chesapeake Bay—Hanging in the Balance." *National Geographic*, June 1993, 2–35.

Hughes, J. Donald. *Ecology in Ancient Civilizations*. Santa Fe: University of New Mexico Press, 1975.

Hutchins, Ross E. *The Amazing Seeds*. New York: Mead & Co., 1960.

International Society for Ecological Economics Newsletter. PO Box 1589, Solomons, MD 20688.

Jenkins, Nancy Harmon. "Urban Cases: A Host's Guide to Farmers Markets." *New York Times Magazine*, 3 June 1990.

Knipe, Tom. "Mass Appeal: Building Steel-Belted Houses," *Calypso Log*, October 1990, 14–17.

Kofalk, Harriet. "Solar Cooking." *Talking Leaves*, Summer 1992, 3–5.

The Land Report. 2440 E. Water Well Rd., Salina, KS 67401.

Lemonick, Michael D. "The Big Green Payoff." *Time*, June 1992, 62–64.

Leopold, Aldo. *A Sand County Almanac*. New York: Oxford University Press, 1966.

Logan, William Bryant. "Turn Garbage Waste into Garden Gold." *Organic Gardening*, July–August 1991, 46–50.

Madson, John. *Where the Sky Began: Land of the Tallgrass Prairie*. Boston: Houghton Mifflin, 1982.

Margolis, Lynn, and Dorian Sagan. *Micro-Cosmos*. New York: Summit Books, 1986.

McCosh, Dan, and Dennis Normile. "Electric Vehicles Only." *Popular Science*, May 1991, 76–83.

Meadows, Donella and Dennis L. *Beyond The Limits*. Post Mills, VT: Chelsea Green Publishing Co., 1992.

Milne, David. *The Evolution of Complex and Higher Organisms*. NASA: Ames Research Center, 1985.

Morse, Mary. "Sustainable Development in Action." *Utne Reader*, Jan.–Feb. 1991, 19–21.

Moskowitz, Robert. "Telecommuting Continues to Gain in Popularity." *Micro Times*, 28 June 1993, 43–48.

"No Sweat: Stay Cool and Save Billions." *Rocky Mountain Institute*, Summer 1992, 1–3.

Nulty, Peter. "Recycling Becomes a Big Business." *Fortune*, August 1990, 81–85.

Odum, Eugene P. *Fundamentals of Ecology*. Philadelphia: WB Saunders Company, 1961.

Pacific Gas & Electric Company, San Francisco. Interviews by Roberta Swan, 1991–1992.

Perry, Nicolette. *Symbiosis*. Dorset, England: Blandford Books Ltd., 1983.

Pohig, James. *Leaves: Their Amazing Lives and Strange Behavior*. New York: Holt, Rinehart and Winston, 1971.

Real Goods Catalog. Willits, CA, Fall 1992.

Rosenfeld, Albert. "Master Molecule, Heal Thyself." *NSF Mosaic Reader, DNA, The Master Molecule*. Wayne, NJ: Avery Publishing Group, 1983.

Sandred, Kjell B. *Leaves*. New York: Crown Publishers, 1985.

Sears, Paul. *Deserts on the March*. Tulsa: University of Oklahoma Press, 1959.

Seidman, Peter. "Solar Car Team Excels Down Under." *Michigan Today*, December 1990, 14–15.

Smith, Emily T. "The Next Trick for Business: Taking a Cue from Business." *Business Week*, 11 May 1992, 68–75.

Sunelco Catalog. Hamilton, MT, Eighth Edition 1992–1993.

Sunworld. Orinda, CA, International Solar Energy Society.

Swan, James A. *Nature as Teacher and Healer*. New York: Villard Books, 1993.

McDonald, Susan, ed. "Superwindow Technology Leads to Super Savings." *E-Notes, Quarterly Newsletter for the International Institute for Energy Conservation*, November 1992, 1–5.

Texas Instruments Press Kit. April 1993.

Tibbs, Hardin B. C. "Industrial Ecology: An Environmental Agenda for Industry." *Whole Earth Review,* Winter 1992, 4–19.

Tranet A Digest on the Alternative and Transformational Movements. Box 567, Rangeley ME 04970.

Van der Ryn, Sim. *The Toilet Papers.* Santa Barbara: Capra Press, 1978.

Yakutchik, Maryalice. "Every Garbage Bag Tells a Story." *USA Weekend,* 24 April 1993, 18.

Zinn, Laura. "Whales, Human Rights, Rain Forests and the Heady Smell of Profits." *Business Week,* 15 July 1991, 114–116.